Policy Debates on Reprogenetics

Published with support of Siblings Boehringer Ingelheim Foundation for the Arts (Geschwister Boehringer Ingelheim Stiftung für Geisteswissenschaften), Heidesheim, Germany

Svea Luise Herrmann, Dr. phil., is currently a researcher at the Institute for Political Science at the University of Hannover, Germany.

Svea Luise Herrmann

Policy Debates on Reprogenetics

The Problematisation of New Research in Great Britain and Germany

Campus Verlag
Frankfurt/New York

Bibliographic Information published by the Deutsche Nationalbibliothek.
Die Deutsche Nationalbibliothek lists this publication in the Deutsche Nationalbibliografie; detailed bibliographic data are available in the Internet at http://dnb.d-nb.de.
ISBN 978-3-593-38792-5

Cover illustration: Genetic analysis © istockphoto/dra_schwartz
Cover design: Campus Verlag

Printed on acid free paper.
Printed in Germany

For further information:
www.campus.de
www.press.uchicago.edu

Contents

Acknowledgments

From the very beginning many individuals were involved in the preparation and realisation of this book in one or another way—to all of them I would like to express my deepest gratitude.

I want to particularly thank Professor Kathrin Braun, who promoted and nurtured my academic activities almost from the start, for her critical, caring, and thoughtful supervision of this study. Her inspiring comments in many discussions and her assistance in structuring sometimes confused thoughts were of great value. My special thanks also go to Professor Barbara Duden for constantly insisting that I reconsider theses and concepts and for often preventing me from getting caught in the discourse trap.

Likewise, the fruitful and productive discussions in the study group on biopolitics at Leibniz University and in the research project EGo helped to further develop and diversify my ideas on the topic. I am particularly indebted to Sabine Könninger, Dr Isabella Jordan, Dr Helen Kohlen, Dr Marion Schumann, and Dr Alfred Moore for their cooperative support and critical feedback. My special thanks also go to Sarah Sexton and Dr Susanne Schultz, who supported my work in different ways, as well as to Dr Christina Oberstebrink for doing a great job in editing this book.

My family was of prime importance with their continuous encouragement throughout my research and writing the book: The loving support of my mother Maria Herrmann, my sister Anne-Kari Herrmann, and my partner Hilke Rusch was invaluable for recuperation and physical and psychological well-being. This book is dedicated to them.

I also thank the Heinrich Böll Stiftung Berlin, who funded my research with a scholarship, as well as the Geschwister Boehringer Ingelheim Stiftung for bearing part of the printing costs.

Hanover, March 2009

Introduction

In December 2000, the British Parliament decided to amend the Human Fertilisation and Embryology Act, allowing so-called 'therapeutic' cloning[1] of human embryos. Roughly a year later, in January 2002, the German Parliament introduced the Stem Cell Act, permitting and regulating import of and research on human embryonic stem cells. Both decisions were the result of year-long public debate ensuing the development of and publications on new reprogenetic practices: the cloning of a sheep in Great Britain and demanding permission to perform pre-implantation genetic diagnosis on human embryos as well as import human embryonic stem cells for research purposes a few years later in Germany. In both countries, public debates arose on whether or not existing law allowed the new practices and whether or not existing law should be revised. In the end this led to more flexible and permissive regulation of reprogenetic science and practices.

Reprogenetics and, more generally, biomedicine belong to the most contested areas of science and technology in Western societies. The notion of 'reprogenetics' refers to genetic-technological practices that combine genetic and reproductive technologies. It is a field of great political conflict and public contestation. Practices such as pre-implantation genetics diagnosis (PGD), embryonic stem cell research, or research cloning give rise to intense public/political debates in many Western countries including Great Britain and Germany (cf. Gaskell/Bauer 2001). Increased public interest in reprogenetic and biomedical developments is reflected in the emergence of political institutions, NGO's, or individuals concerned with reprogenetics regulation and also in an upsurge of more or less institutionalised proce-

1 'Therapeutic cloning' is the transfer of a body cell nucleus into a de-nucleated egg cell. Its target is the production and derivation of stem cells, which are supposed to be 'genetically identical' to those of the person from whom the body cell came. This 'breakthrough' was presented as potentially offering treatment possibilities for diseases linked to so-called cell aberrations.

dures of deliberation between political officials, science, and the public that are associated with biomedical policy in one way or another (Braun et al. 2002; for an overview of European institutions cf. Fuchs 2001; Klüver et al. 2000).[2]

For political decision-making, the issue area of biomedicine and reprogenetics presents "wicked problems" (Fischer 2000, 128, cf. Rittel/Webber 1973), that is to say, problems "in which we not only don't know the solution but are not even sure what the problem is" (Fischer 2000, 128). The very understanding of 'the problems' inherent in biomedical practices is highly controversial. "Wicked problems", as Fischer defines them, cannot be unambiguously or conclusively defined, and, thus, there are no clear-cut criteria by which 'resolution' thereof can be judged. Pluralistic societies are characterised by a multiplicity of values and normative disagreements, which lead not only to conflicts in terms of how to deal with problematic situations or issues but also to differences in regard to the very understanding of what the problem 'really' is. Especially in biomedicine and reprogenetics, it remains undecided whether the problem is one of weighing up 'chances and risks' or whether it is a conflict arising either between 'moral taboos and economic interest' or between 'individual rights and social consequences'. Either way, the definition of a problem motivates the search for a 'solution' and making decisions. Because of the varying comprehensions and definitions of what different actors see as problematic in biomedical or reprogenetic issues, it makes more sense to use the term *problematisations* (Bacchi 1999) rather than problems in this context.

In the case of biomedicine or reprogenetics policy (as well as, probably, science policy more generally), we can speak of a situation of "radical uncertainty" (Hajer/Wagenaar 2003, 9) with regard not only to the foundation and results of biomedical research and development but also to the foundation of decision-making and outcomes of policy-making. As Hajer and Wagenaar (ibid.) maintain, this means that policy-makers cannot make absolute judgements on the grounds of something like 'appropriate knowledge for policy', not least because knowledge foundation underlying decision-making is itself marked by uncertainty—even at the level of science. In the field of reprogenetics, not even distinguished experts agree on the (hoped for) techno-scientific results of research: Indeed, it is also scientific-

2 For an overview on embryo research regulation across Europe, cf. Survey on the National Regulations in the European Union Regarding Research on Human Embryos; http://europa.eu.int/comm/european_group_ethics/docs/nat_reg.pdf, Apr 4, 2002.

ally unclear if embryonic stem cell research will actually lead to 'therapeutic results', what the 'side-effects' of particular procedures may be, or whether adult stem cell research might be a 'better or worse' way to achieve the intended results. Thus, in a situation of radical uncertainty, problems "require us to make 'hard' decisions with only 'soft' evidence" (ibid., 9). Hajer and Wagenaar emphasise that this is not in fact a *new* phenomenon related to the enhanced complexity of society or merely to 'new scientific developments', but it is rather related to the "demise of the myth of absolute knowledge" (ibid., 10). Great conflicts concerning science and technology—such as those concerning nuclear power in the 1970s and 1980s or genetically modified crops and food, and the BSE debacle—and the failure of science as well as politics to prevent the negative consequences of scientific endeavour have created public unease about whether scientific development can be effectively regulated or not. As Fischer (2003b) asserts, this is not only true in the field of science policy but in policy-making more generally: in face of failing to solve problems effectively, such as poverty or unemployment, a technocratic "politics of expertise" (Fischer/Forester 1993b) has become disputable. Science and expertise no longer have a reputation for providing objective and unbiased knowledge, that is, knowledge that is derived independently of interests and power configurations, which escapes moral and social influences. Therefore, adherence to scientific knowledge is "no longer a credible policy-making strategy" (Hajer/Wagenaar 2003, 10). Today, objective facts and values are no longer seen as two distinctly separate worlds, and science is not seen as value-free. There is a growing distrust of science and scientists. This is even more so in a policy area in which science itself is problematised, such as in biomedicine and reprogenetics.

Presenting very complex problems, biomedical and reprogenetic developments challenge political decision-making as development occurs at a fast rate, often escaping public notice. In order to comprehend developments, a highly specialised knowledge is required, which needs to be adjusted and re-adjusted according to rapid changes. Thus, in order to deal with questions concerning biomedical development, decision-makers have to fall back on experts in many ways. At the same time, the complexity of techno-scientific development implies an impact on many different levels, such as social, legal, ethical, economical, cultural, political, or health-related effects, which overlap and cannot be strictly separated from each other.

What is more, also the availability of language suitable to grasp biomedical innovation is limited and inevitably normatively charged: ordinary language is often inappropriate for specifying biomedical innovations (do embryos in-vitro have 'parents' or 'siblings'), so a vocabulary must be found in order to define and tackle the issues. However, names, categories, and concepts have normative implications, because language is not value-free: terms that refer to an embryo, such as 'a cluster of cells' or a 'future child', are by no means purely objective, but imply particular, subjective beliefs and values (Braun 2002; Braun et al. 2002). Already the term 'embryo' has normative implications, as feminists have pointed out (cf. Duden 1993). Likewise, in the issue area under study, actors disagree whether we should speak of 'therapeutic' or 'research cloning' or whether we should discard the term 'cloning' for 'cell nuclear replacement' (cf. Sexton 2001), all of which have different normative connotations and evoke different associations.

To complicate things even more, also the normative foundations for evaluating and judging scientific endeavour are highly contested: Do embryos have the right to 'respect'? Do they have human dignity? Or, more generally: Do we want to restrict techno-scientific developments or should we leave science policy to science and scientists? Such problems are paradigmatic for questions that are highly contested in the public sphere. And, last but not least, the stakes are high: biomedical research and developments seem to promise huge profits, economic growth, and competitive strength in a global research field, thus generating pressures on governments to create a stable environment for research (Bauer et al. Gaskell 1998; cf. also Irwin 2006), while there is no agreement on the issue of whether or not economic goals should be integrated in policy-decisions in the field of reprogenetics. Thus, in regard to the wickedness of reprogenetic or biomedical issues, policy-making faces great uncertainties concerning socio-political-economic outcomes and public reactions. Indeed, policy-makers are confronted with unpredictable research results as well as unforeseeable societal responses to and impacts of the decisions they make.

Additionally, at the theoretical as well as at the empirical level, we see that discourse becomes more salient in public policy and policy analysis (cf. e.g. Braun/Herrmann 2000, Fairclough 1992; Hajer/Wagenaar 2003; Irwin 2006). Particularly in the issue area of biomedicine or reprogenetics, we observe an increase in discursive practices that marks a transformation in political practice. Public discourses and especially *ethics* discourses play an important role therein. From an idealized viewpoint, public participation in reprogenetics policy discourse is a means of democratizing policy-making and securing societal participation in the shaping of reprogenetics policy. Observed critically, however, public discourse is much more ambivalent: As empirical analysis shows, commitment to public engagement in discourse does not exclude a similar commitment to scientific 'progress', but, instead, public ethical discourses prevent rather than allow a more general political scrutiny of the character and direction of scientific development, so that *stimulation of* (public) *discourse* could stabilize rather than reduce power configurations within science policy processes.

The present study analyses and compares the policy discourses on reprogenetic research and practices that took place in Germany and Great Britain between the mid-1990s and the end of 2000/beginning of 2002, focusing on the question of how different actors problematised the issue, how problematisations changed throughout the years, as well as what the particular outcomes were. The empirical analysis identified landmarks related to the emergence of problematisations of existing policy on embryo research and the final change of policy, that is, the new regulatory decisions in terms of new laws. Whilst identifying the 'beginning' of a policy process is largely a superficial endeavour as 'new' policy discourses always build on existing ones, certain events can still be identified, which apparently prompted 'new' policy debates and finally led to a transformation of policy. The birth of the cloned sheep Dolly in Great Britain and the first application submitted by two medical doctors for permission to carry out a preimplantation genetic diagnosis in Germany, as well as a funding application made by stem cell researchers for importing human embryonic stem cells a few years later were landmarks. The significance of these events was not simply due to the fact that they marked scientific 'progress' and 'called' for new policy, but because they received a great deal of publicity: To a greater degree in Germany than in Britain, they were widely and controversially debated in the public sphere as well as by high-ranking policy officials and, indeed, reached the status of policy problems. Policy decisions, that is, the implementation of a new law in Germany allowing and regulating imports of embryonic stem cells and the change of law in Great Britain allowing research cloning (GB), mark the end of the policy debates, despite the fact that both were reason for more controversy. However, implementation of new regulations was taken as caesura for data collection and analysis.

Great Britain and Germany are, scientifically as well as economically, both prominent in the international arena of biotechnology. They are, however, also competitors in this field, a fact that definitely plays a role within the national policy-debates. For example, on the one hand, Tony Blair (as prime minister of the UK) emphasised German competition as a reason to broaden the scope of embryo research, which, on the other hand, provoked Gerhard Schröder (as German chancellor) to stress that the German position in the world biotech-economy must be maintained. Also, in international debates on embryo research at the time, the regulations of both countries often served as 'good' (or bad) examples for different factions: the British Human Fertilisation and Embryology Act (1990, HFE Act) is often considered to be the most comprehensive legal framework concerning embryo research while, at the same time, allowing a broad scope of research, and, as such, is regarded as exemplary; in contrast, the German Embryo Protection Act (1990, EschG) was deemed to be one of the most restrictive of its kind and, as such, was regarded as paradigmatic by British opponents to embryo research, 'proving' that restrictive laws would not necessarily lead to economic or scientific 'decline'.

Thus, while both countries had policies regulating embryo research, these differed as regards comprehensiveness and restrictiveness: while the UK law provided a high degree of experimental freedom on the basis of a system of case-by-case licensing and a law regulating many different aspects of the research, the Embryo Protection Act is part of the German penal law and prohibits embryo research altogether. Both policies came under attack at about the same time because of the development of and the request to implement new techno-scientific practices, and both policy debates ended with a more permissive regulation of scientific endeavour—that is to say, they allowed a wider range of research in the Human Fertilisation and Embryology (Research Purposes) Act 2001 in the UK and permitted imports of and research on human embryonic stem cells in the 2002 new Stem Cell Act in Germany.

While the two policy debates led to a more permissive regulation of science, there are several points in which they to some extent differed. First of all, the legal-normative frameworks vary in the two countries: most obviously in regard to the fact that the British HFE Act already allowed embryo research, whereas the German EschG prohibited it. Additionally, Germany has a written Constitution that is the supreme legal framework for implementation as well as interpretation of law, and which, as such, restricts law-making to a certain extent. In contrast, Great Britain does not have a central constitutional document. The British Constitution consists of written sources, which are technically equal to ordinary law and can be reformed simply by parliament passing an act and unwritten conventions, which are difficult to categorize in a legal sense and not enforceable by courts. Apart from legal differences, also at the societal level there was at least one striking dissimilarity related to biomedical developments in particular: Although in Germany general protest against new biomedical practices and against bioethics was apparent and on a large scale even before the debates under scrutiny began, this was not the case in Great Britain. In spite of the fact that the respective reprogenetics policy debates in both countries began against quite different backgrounds, in both cases the result was that regulation of science was liberalised.

As was mentioned above concerning policy-making, reprogenetics is a "wicked problem" (Fischer 2000), that is to say, there is neither a clear-cut and unambiguous definition of what the problem is nor are there clear-cut criteria for its resolution. Thus, it makes a significant difference how various actors define the issues at hand and which *problematisations* appeared in and dominated policy debates. Furthermore, policy discourses are embedded in the socio-political context in which they appear and thus in social power configurations. Discourse is conducted by actors and actors living in a world of social hierarchies and power. In this connection, the most productive approaches to public policy discourses or, more generally, socio-political discourse are those scrutinizing power effects *on* as well as *within* policy discourse. This is the case with, for example, Fischer and Forrester (1993) or Nancy Fraser (1989a), who claim that discourse is an area of struggle and contestation in which socio-historical factors, such as power relations, social norms and standards, or classification systems, play a decisive role. However, in a critique of these approaches that also extends beyond them, a more general analysis of what I call *discourse stimulation* is conducive to understanding the function and effects of the increase and proliferation of public discourses, as we can especially observe in the issue area of biomedicine or reprogenetics. Thus, at a more general level, this study adopts a notion of discourse based on that developed by Michel Foucault: Discourse is comprehended here as a powerful and power-laden social activity, which not only has a semantic dimension (i.e. the production of meaning) but also a practical one, that is, the production of discourse and integration of actors into a power nexus. Thus, besides a semantic analysis of problematisation, the issue of discourse stimulation in reference to (but also going beyond) Foucault's discussion of the productive qualities of discourse is to be scrutinised. In this way, not only discourses but also actors' activities are considered. The questions are: Who initiated policy discourse? Who presented which problematisation? What are the implications and effects of discourse stimulation?

Ethnographic science, technology, and society (ST&S) studies provide a further theoretico-analytical framework that is useful for the analysis of reprogenetics policy discourse. ST&S studies are engaged with the governance of science and the (problematic) relationship between science and society. They also concentrate on questions of framing, especially on the effects of scientific frames dominating in science policy, which are based on a de-contextualised definition of scientific knowledge as 'objective' and neutral. In order to counteract scientistic policy-making, ST&S studies demand that policy processes do not exclusively involve scientific knowledge and scientific actors. Lately, ethnographic ST&S studies also critically assessed the 'success' of their demand for inclusion and participation, concluding that the commitment to scientific 'progress' has survived new participatory modes of science policy seriously constraining the scope for dialogue. The questions ensuing from the normative demand for participation as well as the results of critical analyses of its success are: What role did the assumption of 'sound science' play in problematisation processes and whether or not public debates offer the possibility of challenging this assumption?

Altogether, techno-scientific developments should be open to *societal political shaping*, also including questioning if scientific endeavour and 'progress' are desirable, because science policy should not be formulated in professional circles alone. Indeed, public policy debates should principally allow for questioning of scientific developments and the outcome of the former remain open. Thus, in general, it must be asked if public policy debates on reprogenetics can have a serious impact on the political shaping of scientific development as well as which role problematisation processes and processes of discourse stimulation play in this respect.

Regarding analysis of policy discourses in the area of reprogenetics in Great Britain and Germany, different theoretical approaches to science policy-making and, more generally, public policy offer fruitful insights and will be introduced in Part 1: The first chapter presents post-positivist approaches to public policy analysis. Interpretive approaches to public policy have shown that policy-making is not simply the positivist search for the objective truth of a problem and its solution, but rather a discursive process of problem formation and argumentation. Interpretive approaches to public policy present theoretico-analytical tools for comprehending the processes of problematisation and their impact on policy-making. In this respect, it is essential to understand the policy process as a discursive process: Rather than assuming the existence of objective, 'pre-set' policy problems, interpretive policy analysis understands policy discourse as a struggle over the meaning and definition of policy problems. Most discourse-analytical policy analysis draws on Michel Foucault's notion of discourse, wherein discourse is a power-laden social practice and a site for power struggles. In this chapter, the notions of *problematisation*, *framing*, and, more generally, *discourse* are introduced and elaborated upon. Additionally, this part discusses the concept of *discourse stimulation*, or, more precisely, the invitation to discourse as a means of power-production.

The second chapter of Part 1 views investigations of the governance of science and technology. Among the several authors in the area of science, technology, and society studies (ST&S), especially those connected to the ethnographic turn in studies on the relationship between science and society were engaged in and formulated critiques of the analysis of what is called the public understanding of science (PUS) and a deficit model of the public, that is, an understanding of public knowledge of science as being uninformed and requiring education. In contrast to traditional PUS, which is based on the assumption of the authority and objectivity of scientific knowledge, the authors of ethnographic ST&S studies have developed an understanding of knowledge as culturally embedded and, more importantly, they have developed an understanding of diverse knowledges as being equally intellectually sophisticated and legitimate. More recently, research has likewise focused on the 'success' of their critique, namely, on what appears, on the surface, to be a decline of the deficit model and a shift toward not only more (public) participation, but also toward integration of forms of knowledge other than scientific knowledge within policy-making. ST&S critique of the dominance of scientific epistemology, in addition to its impact on 'new' participatory modes of science government, enhances the critical analysis of science policy-making, including reprogenetics policy, by an increasing acknowledgement of knowledges other than scientific knowledge and of non-scientific participants in policy discourses. Indeed, research findings in this area are particularly helpful for analysis of the power-effects of what I have termed discourse stimulation.

The following part presents the empirical analysis of policy discourses on reprogenetics in Great Britain and Germany. Each study of the respective country begins with a short introduction to the historical background of the individual debates. The focus is especially on their genesis as well as the content of the legal instruments that came under attack—that is, the British Human Fertilisation and Embryology Act and the German Embryo Protection Act—and the particular discursive backgrounds, with a brief resume of the main lines of debate and conflict that preceded the period of study. This will be followed by an analysis of the respective policy discourses in both countries. The conclusion of this section will compare processes of problematisation and discourse stimulation and then draw conclusions on the issue of how framing and problematisation affected policy processes, what impact they had on policy outcomes, as well what role stimulation of discourse played. A particular focus lies on the question of whether or not and in what ways problematisation and discourse stimulation promoted or impeded the societal political shaping of scientific developments.

Theoretical Framework

1 Public Policy Analysis: How to Approach Reprogenetics Policy Discourse?

"Public policy analysis often appears as a jungle of diverse and conflicting modes of inquiry, full of inconsistent terminologies, divergent intellectual styles—perhaps indeed incommensurable paradigms." Douglas Torgerson (1986, 33)

The notion of *public policy* is by no means a straightforward or agreed concept. Some authors understand public policy to be "what governments choose to do or not to do" (Dye, cited in Fischer 2003b, 2); others identify it simply with lawmaking or prescriptions backed by sanctions (cf. Fischer 2003b, 2). Schubert and Bandelow (2003, 4) define 'policy' as "the substantial, material dimension of politics", while Heclo (as quoted in Parsons 1999, 13) emphasises the "purposive" dimension of policy decisions.

Douglas Torgerson (1986) distinguishes "three faces of policy analysis", stating that they differ in diverse perspectives on the relationship between knowledge and politics. Historically, as Torgerson (ibid., 34) writes, the emergence of the three faces relates to developments in the social sciences associated with positivism, the critique thereof and disillusionment with it, and the development of a post-positivist direction.[3] According to Fischer, the advent of "professional policy analysis can be traced back to the progressive movement which called for increasing the role of expertise in government" (2003b, 2) because of the growing need for advice concerning political decision-making in an increasingly technologically complex society. The emergence and development of policy science must be considered in the context of the "rationalisation of the state and politics"

3 The separation and historical sequence of the three faces, however, is certainly artificial and historically misleading. Rather, they overlap and impregnate each other (Torgerson 1986, 34).

(Parsons 1999, 16) and the "desire for knowledgeable governance, that is, the acquisition of 'knowledge' about 'problems' so as to formulate 'better solutions'" (ibid., 17). The positivist approach to policy seeks to abolish politics in favour of orderly and knowledgeable administration, and the mode of positivist policy analysis is to *collect* knowledge *of* society in order to *apply* it *to* society (Torgerson 1986, 35). In this respect, natural scientific methods serve as a model for a rationalistic policy approach, which comprehends political issues as technical issues requiring administrative management, such as collecting 'objective' data, applying technical methods (for example, risk assessment or calculation), in order to find and implement (technical) solutions (Fischer 2003b, 5). The positivist policy approach is not only based on confidence in the knowledge to be gained but also on a particular concept of knowledge: 'real' knowledge is scientific knowledge—objectively observed facts that are characterised by regularities and can be used to predict future events and employed in the development of a social technology (Torgerson 1986, 36). The first face of policy analysis—associated with positivism—strictly separats facts from values and policy analysis is seen as an "adjunct to the necessary unfolding of a benign and objective pattern of development—the orderly progress of a rational civilization" (ibid., 37).

The second face, as Torgerson describes it, appeared when policy analysts realised that they were "blinded to political reality" (ibid., 37) and had failed to see the context of policy, the "hidden political dimension", (ibid. 38) that, for example, serves particular interests or reinforces the order of an established political world. However, despite the recognition of the "hidden political dimension", old positivist notions of science's ability to deliver solutions persisted. The critique of the first face was a critique of the unavoidable political 'misuse of knowledge' rather than of positivist understanding of knowledge. In summary, "in the first face of policy analysis, knowledge replaced politics; in the second face, politics overwhelmed knowledge" (ibid., 39).

Often called the founder of the policy approach (Fischer 2003b, 4; Parsons 1999, 21; Torgerson 1985), Harold Lasswell was the first to differentiate between knowledge *in* and knowledge *of* the policy process (Lasswell 1970). His goal was a multidisciplinary approach to regulate political decision-making processes. He thereby acquired the reputation of being a positivist, although his approach was "clearly at odds with positivist posture" (Torgerson 1985, 242). Lasswell promoted a contextual orientation in social science that could mediate between academics, decision-makers, and the public in order to minimise conflicts on pressing policy issues (Fischer 2003b, 3; Torgerson 1985, 242). He emphasised the "inevitable participation of the inquirer in a human world" (Torgerson 1986, 42) as well as citizen participation (ibid. 43). Lasswell's normative goal was a 'policy science for democracy'. However, according to Fischer (2003b, 3), policy science failed to follow Lasswell's vision of a policy science for democracy. Instead, policy science "emerged more along technocratic rather than democratic lines" (ibid., 3f) and was shaped by neo-positivist methods dominating the social sciences at the time:

> "This has generated an emphasis on rigorous quantitative analysis, the objective separation of facts and values, and the search for generalizable findings whose validity would be independent of the particular social context from which they were drawn: that is, a policy science that would be able to develop generalizable rules applicable to a range of problems and contexts" (Fischer 2003b, 4).

The third face Torgerson identified is of the emerging post-positivist policy analysis in which "we glimpse at the potential for a relationship in which politics and knowledge are no longer deadly antagonists" (Torgerson 1986, 39). This third face of policy analysis, or rather the ways it developed after Torgerson wrote his article, is the subject of the following chapter.

1.1 Interpretive Approaches to Policy Discourse

> "In actual practice, policy analysts and planners do a great deal more than they have been given credit for doing." Frank Fischer and John Forester (1993b, 2)

Post-positivist approaches to policy analysis draw attention to the fact that policy-making does not simply respond to pre-set policy problems. They counter the positivist view that reality, truth, and knowledge are the results of objective rules as well as the view that problems can be explained and solved via the application of (factual) knowledge. For positivist theory "[p]roblems were 'facts', they could be measured and dealt with in a 'scientific' manner: to analyse causes was to solve problems" (Parsons 1999, 95). The positivist approach to public policy assumes that there is an accessible objective real world and that, correspondingly, the policy process can be set out in clear-cut stages of problem identification, rational approach to the problem, collection of factual data, as well as policy formulation and implementation. A fundamental positivist principle is to separate facts from values, that is to say, (empirical) research must function independently of context as well as normative implications in order to be judged as scientifically valid research (Fischer 2003a, 212).[4] According to Fischer, policy sciences attempting to adhere to this principle have endeavoured to translate normative issues into technically defined goals attainable by technocratic administrative means. Positivist policy science understood problems as issues that demanded managerial and programming improvement via formulating goals and objectives, determining probabilities, and collecting objective data (Fischer 2003b, 4). Emphasis of the need for scientific competences in the decision-making process and taking recourse to expertise is intrinsic to this understanding of problems and policy-making.

> "Associated with this belief in the superiority of scientific decision-making has been an emphasis on the calculation of the efficiency and effectiveness of means to achieve goals, with the result that policy analysis has in significant measures evolved as a strategy of 'technocratic form' of governance" (ibid., 5).

As Hajer/Wagenaar (2003, 6) state, positivism is, first and foremost, a particular attitude towards knowledge that implies not only a particular factual understanding of reality but also methodological principles of data collection and norms of articulating knowledge. In practice, this means that "positivism includes normative beliefs and habits of governance and policymaking" (ibid.). In opposition to the positivist view of facts as the ultimate device for solving societal problems and fact-finding as a means of shaping an adequate policy, post-positive approaches question the belief in an objective science for establishing objective truths and finding, in this way, solutions. What appears as 'fact' in the sense of a concrete and incontestable reality in positivist theory is comprehended by post-positivist scholars as being based on interpretations that involve choices and judgements (Bacchi 1999, 1) and hence inherently, although not obviously, normative. For post-positivist scholars it is not only impossible to find 'objective facts' or a neutral vocabulary to describe them, but they also state that rationalistic policy analysis serves an ideology that masks elite political interests (Fischer 2003b, 14). As Fischer writes, (neo-)positivist policy analyses seek to transform emotional and conflict-ridden political ques-

4 Fischer (2003a) discusses the catches and misunderstandings of positivism or, in his words, "empiricism". He gives an overview on the emergence and development of policy studies and confronts empiricist/positivist and interpretive/deliberative policy analysis.

tions into scientific ones, addressing them technically and searching for technocratic managerial solutions. Thus, in Fischer's words, they serve "to facilitate and bolster bureaucratic governance" (ibid.).

In contrast, the notion of *interpretive public policy* implies both content- and process-related aspects of politics. It refers to the process of problem-analysis and -definition, agenda setting, development and selection of solutions, policy-design, as well as implementation and evaluation of policies. It deals with the questions of how problems are presented or interpreted by different actors as well as how interpretations of problems differ and change throughout the policy process. As an academic discipline, interpretive policy analysis explains and evaluates policy processes while also examining methods of investigation. Very generally speaking, interpretive policy studies inquire into why and how issues come to be defined and constructed as problems and how they are placed on the political agenda. It scrutinizes different actors, their discourses, argumentative strategies, as well as naming and framing processes (cf. Bandelow/Schubert 2003; Parsons 1999, xv).[5] While many post-positive policy analysts are involved in both, we must distinguish between policy analysis *of* the process and policy analysis *for* the process (Lasswell 1970). While the first is an *academic* discipline directed at understanding how policy-making functions, the latter refers to the *practice* of analysis within the policy process. It must be noted that this study focuses on comprehension *of* the policy processes on reprogenetics in Germany and Great Britain and does not intend to recommend ways of improving them. Thus, the approaches introduced will serve as analytical tools for the investigation of policy discourses on reprogenetics in the two case studies.

Problem Construction and Values

Initiated to a great extent by US-American policy academics, the so-called "argumentative turn in policy analysis and planning" (Fischer/Forester 1993a) designated a shift within policy analysis towards the wish to understand "just what policy analysts and planners do, how language and modes of representation both enable and constrain their work, how their practical rhetoric depicts and selects, describes and characterizes, includes and excludes, and more" (Fischer/Forester 1993b, 2). Interpretive approaches

5 Parsons (1999) and Schubert/Bandelow (2003) provide excellent overviews of the concepts, approaches, and methods of policy analysis.

emphasise the normative impact of public policy and the construction of policy problems: "how we perceive and think about something", Carol Bacchi (1999, 1) states, "will affect what we think ought to be done about it". Despite many differences, post-positivist scholars share the assumption that policy controversies are not mainly controversies of opinion, but that they are due to differences of values, beliefs, and interests. They claim that valuation and judgement are unavoidable. Starting from the assumption that the language we use to talk about policy and problems is not neutral but value-laden, these approaches look at ways of how a problem is defined and framed, and how thereby meaning is produced. Individual actors may interpret a problem differently. However, differences do not simply arise from competing opinions on particular issues but from the fact that political 'problems' are shaped by underlying normative assumptions, beliefs, and values. The post-positivist approach to policy analysis focuses on the insight "that a certain conception of the way a scientific method should proceed, and its grounding in beliefs in epistemology, almost inevitably lead to a certain conception of society; an understanding of how society should be organized and managed" (Hajer/Wagenaar 2003, 13).

In this view put forward by Hajer and Wagenaar, policy controversies are disputes over meaning and value rather than simply a difference in opinion. Thus, practical policy analysis is not merely concerned with setting goals and finding means to achieve them, but the most important function of practical policy analysis and policy-making is to define the determining norms when a certain condition is regarded as a problem (Majone 1989). Carol Bacchi's (1999, 148-163) analysis of the history of the social problem status of abortion illustrates how the framing of an issue, abortion, in terms of normative assumptions and standards, affects whether the issue finds its way onto the agenda as well as how it is represented as a problem. Bacchi discusses how abortion acquired problem status in the US in the sixties and seventies. She rejects the assertion that abortion was problematised because feminist campaigners demanded that it be addressed politically. On the contrary, feminists wished to make abortion a *non*-problem in opposition to those who were against abortion and demanded that it be a criminal offence. Likewise, medical professionals did not want abortion to be politicised, but rather strove to have abortion, which had become a common practice, regulated in ways that maintained their professional autonomy and did not criminalise it. Bacchi (ibid., 152) concludes that changes in abortion law were less a response to feminists' demands than to the arguments put forward by medical professionals. In *Roe vs. Wade*, although abortion is framed as a private matter, its framing is likewise tied to the "freedom from state interference granted the pregnant woman in the first trimester [...] to her relationship with the doctor" (ibid., 154), including the possibility for doctors to refuse abortion for 'moral' reasons. Catherine MacKinnon (1989) emphasises that framing abortion as a *privacy* issue had negative effects for women, as this not only meant that *public* funding of abortion could easily be withdrawn. It also entailed reinforcing a separation of public and private, in which the private appeared as a sphere that was *per se* free from power, with the effect that *private* and power-laden relationships could not be problematised. The framing of abortion in terms of privacy strengthened rather than weakened private power relations as well as concealed the impact of social and political factors on private relationships (ibid.). Additionally, as Bacchi (1999, 159) points out, a medical framing of abortion meant that a woman's 'right to privacy or choice' was at any rate constrained by the need to consult a doctor.

Thus, far from being a neutral or objective endeavour, the policy process is very much a process of norm setting, and those engaged in policy processes and decisions are themselves involved in the norm-setting process: via the politicisation of a problem while leaving out others; via processes of framing a problem in a particular way or suggesting a particular solution etc. Problem constructions necessarily require value judgements or some kind of 'social vision'. They inevitably entail normative consequences in regard to what counts as valid knowledge in the policy process, what is regarded as justified argument, who is allowed into "the halls of decision-making" (Hajer/Wagenaar 2003, 13), or what are the accepted rules of political conduct.

The Argumentative Turn

Argumentative approaches to policy analysis seek to explicitly include normative assumptions and incorporate wider social meanings instead of only considering allegedly 'pure facts'. This approach no longer comprehends the practice of policy analysis (*for* the process) as the production of truth—that is, analysis as science—but as argumentation that refers to and is embedded in a specific politico-social context—that is, analysis as argument (cf. Dryzek 1993; Saretzki 2003). As an applied science interpretative policy analysis can be understood as a response to or acknowledgment of the existence of "wicked" (Fischer/Forester 1993b) or "intractable" (Rein/Schön 1994) policy controversies, that is, of policy controversies stemming from differences in value judgements, beliefs, and interests rather than lack of expertise and knowledge. Within this context, policy analysis (*for* the process) could not fall back on a positivist model of fact-finding and problem-solving, but had to reflect its own status. Nonetheless, as Hajer and Wagenaar (2003, 6f) state, it would be short-sighted to understand the argumentative turn as a mere counter-narrative to positivism. Rather, it is an attempt to reflect on the normative as well as empirical relationship between policy analysis and the political or social environment in which it functions—that is, including an understanding of how different parties frame a situation, what are the normative or social grounds for this, and what are the implications involved (Bacchi 1999). In a process of self-reflection, advocates of the argumentative turn promoted an understanding of policy analysis and planning as *argumentative processes* in opposition to an objective, fact-finding process. They responded, on the one hand, to the changing status of science, in which science was no longer seen as the source of objective knowledge, and, on the other, to a transformation of politics and policy-making itself (cf. Saretzki 2003). Failures and unwanted consequences of rational planning as well as centralised and hierarchical regulation have created an unease among citizens about the appropriateness and effectiveness of traditional politics of expertise (Fischer 2003b; Fischer/Forester 1993a; b; Hajer/Wagenaar 2003), in which a "knowledge elite" (Fischer/Forester 1993a, 22) dominates a technically defined policy process.

In the argumentative approach to public policy, it is not the facts related to problems but the arguments that are the objects of analysis: "Whether in written or in oral form, argument is central in all stages of the policy process" (Majone 1989, 1). According to Fischer/Forrester (1993b, 4), arguments involve at least two aspects: the aspect of analysis and analytical content as well as the aspect of articulation and practical performance. In line with the way Fairclough (1992) describes the dialectic relationship between discourse and social reality (cf. 1.2), scholars of the argumentative turn comprehend argumentation as a communication process taking place in a specific socio-historical context. The context must therefore be analysed. From their point of view, argumentation is always bound to the socio-historical context in which it appears and is always political.

This implies that it is a medium for struggle and contestation in which socio-historical factors, such as power relations, social norms and standards, or classification systems play a decisive role (Fraser 1989a). Hence, the analysis of argumentation involves the analysis of "(extra-discursive) social practices from which social constructs emerge" (Hajer 1993, 45), including the positions that are being countered or criticised. In this view, argumentation is not a singular process of constructing reality, but it is a social practice that entails strategic efforts and persuasion, interests, manipulation, symbolic significance, construction and reconstruction, framing and reframing of the issue, and more.[6] However, Majone (1989, 8) stresses that persuasion/argumentation does not necessarily entail propaganda, manipulation, and dishonesty but is essential also within 'free debate', wherein participants can promote their own interests and views and are, furthermore, encouraged to modify their values and adjust their views.

Understanding the policy process as an argumentative process embedded in a particular social situation also implies an understanding of this process as a complex and (potentially) power-laden exercise. As Fischer and Forester (1993b, 1) put it: "If analysts' ways of representing reality are necessarily selective, they seem as necessarily bound up with relations of power, agenda setting, inclusion and exclusion, selective attention, and neglect." The most important contribution of the argumentative turn to the understanding and performance of public policy analysis is thus "directing our attention to the way in which knowledge and rationality are central to understanding the nature of power in modern (and post-modern) society" (Parsons 1999, 153).

The goal of the argumentative approach is to illuminate power relations and effects in order to educate those involved in the policy process and to democratise the process itself. It searches for more democratic ways of argumentation in which power could be reduced in order to find more legitimate solutions (cf. Fischer 1999; 2000). Despite—or because of—the understanding of policy-making as an argumentative and thus necessarily biased process, the argumentative group is committed to democracy. Hence, scholars promote the democratisation of policy-making and suggest more participation and proceduralisation of the policy process (Saretzki 2003, 409). In regard to what Fischer (2000) called "wicked problems" or Rein/Schön (1993, 145) labelled "stubborn policy controversies"—that is,

---

6 For an overview of the issues that the *argumentative turn* deals with, cf. articles in Fischer and Forrester (1993a).

problems and controversies that go beyond differences of opinion and are based in value or framing differences—their goal is to find ways of explaining and resolving them in a democratic way (cf. Hajer 1993; Rein/ Schön 1993) by displaying framing differences as well as power structures.

For the analysis of policy processes and policy-making in the field of reprogenetics, the interpretative approach to public policy has the advantage of understanding policy problems not as objective representations but as *interpretations* of 'reality'. This leaves room in which to consider different existing definitions of what the problem 'is', without necessarily stratifying them by assuming them to be 'correct' or 'false' definitions. At the same time, this approach is alert to power configurations, normative assumptions, as well as visions underlying problematisations in that it adopts a perspective devoted to the social embeddedness of policy discourse. Additionally, and this is important for the policy issue under study, it is also committed to the social embededdness of knowledge and science. Not only does this approach to public policy acknowledge that problematisations are the result of and include normative assumptions and visions in science-based definitions, but it likewise admits that problematisation as well as argumentation are (potentially) power-laden activities. In this respect, a more detailed understanding of *discourse* and *framing processes* as sources and effects of power as well as sites for power struggle is useful.

1.2 Policy Analysis as Discourse Analysis

"The struggle for power is a struggle for setting the discourse in which the problem is framed." Wayne Parsons (1999, 152)

While in the positivist tradition language is seen as a neutral tool to describe reality, post-positivist scholars claim that policy problems do not exist outside of language, values, and beliefs. Post-positive contributions to policy analysis not only question the epistemological separation of fact and value, of objective and normative aspects of policy analysis and policy-making. They also reject the potential of neutral language to formulate a hypothesis and describe observations. In contrast, more recent approaches to public policy draw attention to the fact that policy discourses do not simply describe and respond to pre-set (social) problems, but they claim that policy problems are subject to processes of interpretation and social construction. These approaches focus on the various constructions or interpretations of the problem and on the effect they have on solving problems (for example Bacchi 1999; Edelman 1988), emphasising that policy problems are not simply 'out there' waiting to be solved, but are discursively constructed and reconstructed throughout the policy process itself. The 'problem' is the result of a complex representation process and of competing ways of representing strategies (Gottweis 1998, 21). For these scholars, policy discourses are communicative processes in which certain societal situations are transformed into policy problems that demand solutions. Post-positivism is thus also a sceptical attitude, which recognizes that different discourses, definitions, and questions lead to different policy 'problems' as well policy recommendations. It focuses on the central role of language, discourse, rhetorical argument, and stories in "framing both policy questions and the contextual contours of argumentation, particularly the ways normative presuppositions operate below the surface to structure basic policy definitions and understandings" (Fischer 2003b, 14). Post-positivist scholars want to uncover subjective and often invisible political presuppositions underlying what is presented as objective 'truth'. In this perspective, language is no longer considered a tool for objectively *describing* the world, but it is understood as a medium through which actors attach meaning to and thus *create* the world. Discourse, therefore, has a semantic dimension, in that it creates meaning, and a practical dimension, as it is a communicative practice. Rein and Schön define policy discourse as "the interaction of individuals, interest groups, social movements, and institutions through which problematic situations are converted to policy problems, agendas are set, decisions are made, and actions are taken" (Rein/Schön 1993, 145).

Especially Michel Foucault was decisive for understanding discourse as structuring and creating realities. Language and discourse—or more broadly, bodies of knowledge—are the protagonists in many post-positivist and constructionist policy analyses. As an epistemic system, discourse is understood to have constitutive power that structures social definitions, meanings, and interaction in a socio-cultural system. Marten Hajer (1993, 45) defines discourse as an ensemble of ideas, concepts, and categories that give meaning to a phenomenon. It frames certain problems and provides tools by which problems are constructed. Thus, discourse is not simply a piece of *text*, but at the same time an instance of *discursive* and *social practice* (Fairclough 1992, 4). The discursive approach (including all three dimensions of discourse) emphasises the productive aspects of language: a text is the result of a social and discursive process of interpretation and production and creates (a particular perspective of) the world. Discourse is seen as a mode of action rather than a mere reflection of the world.

The Productive Character of Discourse: Framing and Problematisation

Thus, within post-positive policy analysis, the notion of discourse acquires a key position. As Edelman (1988, 12) writes, "[t]roubling conditions come into discourse and therefore into existence". Likewise, Carol Bacchi (1999) emphasises that language and discourse do not represent reality but, on the contrary, actively produce it: meaning that the object of policy processes, that is, the policy problem, is not merely identified and named but created. "Problem definition", as Rocheford and Cobb (1994, vii) assert, "has to do with what we choose to identify as public issues and how we think and talk about it". Whether or not a troubling condition is regarded as a political or societal problem depends upon how it is discursively constructed, defined, and interpreted as such (Hajer 1993, 44).

The understanding of the positive character of discourse is based on the theory of discourse developed by Michel Foucault. One of the most important and unique claims in Foucault's political thinking is that power and discourses have a positive-productive character. Power, he maintains, is inherent in each and every social relation, institution, and discourse. Foucault (1998), in his analysis of discourses on sexuality in the nineteenth century, elaborates the theory that knowledge and truth are generated through discourses. In opposition to the commonly assumed liberalising function of discourse, Foucault suggested that discourse produced sexuality and did not liberate it. "Truth is a thing of this world" (Foucault 1980, 131), that is, "truth" is not something to be found and named, but something produced within the power-laden order of discourse. In the words of Michel Foucault, discourses are "practices that systematically form the objects of which they speak" (Foucault 1989, 54). They "do not identify objects, they constitute them and *in the practise of doing so conceal their own intervention*" (Foucault 1989, 54, emphasis added). The important insight adopted from Foucault is that objects of discourse are regarded as being constituted and transformed within discourse rather than existing independently. In the Foucauldian sense, 'objects' are "objects of knowledge, the entities which particular disciplines or sciences recognize within their fields of interest, and which they take as targets for investigations"

(Fairclough 1992, 41). Hence, with regard to the productive character of policy discourse, Carol Bacchi (1999, 2) suggests that it is more apt to speak of *problematisations* rather than problems, emphasising that 'problems' acquire their shape, definition, and meaning through discursive production processes.

Referring to Erving Goffman (1974), Martin Rein and Donald Schön (1994) introduced the notion of *framing* within policy analysis. In their terms, frames can be thought of as stories about a difficult situation in which the author describes the problematic aspects of the situation as well as how it can be resolved. Different frames tell different stories about reality and represent distinct ways of seeing it. Framing, as Rein and Schön define it, is a "way of selecting, organizing, interpreting, and making sense of a complex reality to provide guideposts for knowing, analysing, persuading, and acting. A frame is a perspective from which an amorphous, ill-defined, problematic situation can be made sense of and acted on" (Rein/Schön 1993, 146). At the same time, framing affects ways of understanding a (problematic) situation and directs attention and channels policy discourse and policy-making in a particular direction, without, however, necessarily being used strategically:

> "The name assigned to a problematic terrain focuses attention on certain elements and leads to neglect of others. The organizing of the things named brings them together in a composite whole. The complementary processes of naming and framing socially construct the situation, define what is problematic about it, and suggest what courses of action are appropriate to it. It provides conceptual coherence, a direction for action, a basis for persuasion, and a framework for the collection and analysis of data—order, action, rhetoric, and analysis" (ibid., 153).

In short, frames provide coherence in an incoherent and ill-defined world. It has to be stressed, however, that frames are not simply invented by their proponents, but they are bound and attached to certain values and normative assumptions effective in a socio-cultural environment, which gives them their normative force (Rein/Schön 1994, 27). In this sense they are relative (as opposed to objective or certain): they are relative to the social position and normative orientations of those employing them (Fischer 2003b, 136).[7] As Fischer (ibid., 144) emphasises, frames highlight certain

7 Fischer discusses the 'problem of relativism' and how it was criticised by positivists, who maintained that it did not offer the possibility to clearly test assumptions and conclusions and that relativist arguments were amoral because they could be used to support any case or discussion. Fischer concludes, however, that it is more relevant to the moral

elements of reality at the cost of others and thereby enable certain ways of thinking and talking about issues while disenabling others. This, however, does not suggest that framing can be avoided. On the contrary, "no one is exempt from the need for framing" (Rein/Schön 1993, 147). Hence, the question is not whether framing processes occur, but rather how an issue is framed and what are the implications thereof. Although hidden, the most important aspect of framing processes is that they are not necessarily obviously productive but, instead, their productive character is concealed and the result appears as the problem as such.

Dialectics of Discourse and Social Reality

Norman Fairclough underlines a *dialectic* relationship between discourse and social structure, in which the latter is not merely an effect but also a condition of the former.

> "On the one hand, discourse is shaped and constrained by social structure in the widest sense and at all levels: by class and other social relations at a societal level, by the relations specific to particular institutions such as law or education, by systems of classification, by various norms and conventions of both a discursive and a non-discursive nature, and so forth. [...] On the other hand, discourse is socially constitutive. This is important of Foucault's discussion of the discursive formation of objects, subjects, and concepts. Discourse contributes to the constitution of all those dimensions of social structure which directly or indirectly shape or constrain it: its own norms and conventions, as well as the relations, identities and institutions which lie behind them. Discourse is a practice not of just representing the world but of signifying the world, of constituting and constructing the world in meaning" (Fairclough 1992, 64).

Hence, discourse is neither the only source of the social nor a mere reflection of social reality—but both at the same time. While it certainly has an effect on the construction, reproduction, and transformation of social identities and relations, as well as on systems of knowledge and belief, it is itself also an effect of these social conditions (ibid., 64f; cf. also MacDonald 2003). Discourse, Fairclough insists, inevitably takes place within a material reality, more or less stable power configurations, and preconstituted social subjects or objects. At the same time, discourse interacts

---

context in decision making to recognise that interpretations vary according to social context than a position of certainty is. Indeed, he states that a position of certainty is much more harmful than reflective relativism (Fischer 2003b, 136f).

with these factors and may also attempt to transform them (Fairclough 1992, 60). This dialectical view counters the identification of discourse with social reality and practice: Discourse is not identical with social reality or practice, although they are inseparably bound to each other and interdependent. "Social practice has various orientations—economic, political, cultural, ideological—and discourse may be implicated in all of these without any of them being reducible to discourse" (ibid., 66).

A perspective that exclusively centres on discourses as speech camouflages the material aspects of the social and thus forecloses their critique. For example, access to discursive resources or the ability to partake in discourse can be limited on the basis of different social factors, such as economic situations or social stratification, which cannot be reduced to their discursive dimension alone, as Fraser (1989a) demonstrates. The assertion that the material dimensions of the social are the result of discursive formations does not offer any analytical gain. Instead, it prohibits critique of social relations, material conditions, or social hierarchies. Hence, while stressing the productive quality of (policy) discourse, a social-constructionist approach to policy discourse presumes that constructions are shaped within a social and material context. This means that, although policy problems are considered to be the result of construction processes and not simply reflections of reality, most scholars of the discursive approach to public policy counter the belief that all constructions are equally valid and acceptable simply because there is no objective truth to test them (Bacchi 1999, 38f). On the contrary, most they insist that constructions are bound to social situations, language, and power relations (e.g. ibid.; Fischer/Forester 1993b). As Nancy Fraser (1997b) states, forms of public discourse are bound to societal power structures and hierarchies, which have significant effects on discourse, discursive constructions, and problematisations. Thus, problematisations reflect material and socio-cultural conditions, constraints, and opportunities and vice versa. The interpreter's situation plays a decisive role in the construction process, depending, for example, on authoritative support or identification with a particular group (Edelman 1988, 3; cf. also Ingram/Schneider 1997). Not all possible interpreters have the same resources to initiate a policy debate, not all have the same access to the media to make themselves heard in public, not all have the financial resources nor are in the position to participate in conferences or debates, much less to become a member of an advisory commission, nor the linguistic skills and abilities to speak in public. Nancy Fraser

(1989a) has illustrated that societal power configurations and hierarchies have a significant effect on forms of discourse, including discourse boundaries, endorsed and excluded topics, interpretation of conflicts, discursive patterns, positioning and limits of participants, or the division between public and private, and more. She concludes that public discourses are embedded in and affected by power configurations, which are not simply transcended and neutralised within discourse.

A further, equally important reservation against a view that considers discourse to be the main pre-condition of the social, including social identity of the subjects, is that this perspective rules out active social agency (Fairclough 1992, 45). This view not only excludes the analysis of different ways of responding to and actively intervening in discourses but, indeed, also the scrutiny and analysis of struggles for discourse.

> "Discourse as a political practice is not only a site of power struggle but also a stake in power struggle: discursive practice draws upon conventions which naturalize particular power relations and ideologies, and these conventions themselves, and the ways in which they are articulated, are a focus of struggle." (ibid., 67)

John Dryzek (2002, vi) stresses that discourses are not free from power by definition but structured by inequalities and power relations. Therefore, discourse is a sphere of conflict, which is exactly the reason why discourses necessitate critical reflection from an internal point of view. Provided discourses present the opportunity to critically reflect on and struggle over discourse itself, they have the potential for the reduction of power (ibid.). Correspondingly, Salter and Jones (2003, 25) assert that the "ability to rise to a position of power is partially contingent upon a skilful command of language". However, also the reverse is true: Command over language depends on positions of power. Actors do not simply respond passively to social constructions, but constructions are objects of social processes in which "even the most routinized practice of social agents involves the active construction, deconstruction and reconstruction of social meanings" (Fraser 1989b, 156). The social, states Fraser, is by definition a space of contestation, and politico-social discourses are marked by a competitive plurality of interpreting modes and talking about issues, which are subject to challenge, conflict, and change. There are struggles over the political status of a problem, over its interpretation, and over who has the power to authorise and gain support for their interpretation (ibid.). In effect, interpretations are not stable systems of meaning but may be subject to rapid change. Problematisations—or, more accurately, the dominance of particular problematisations—are thus an effect of the dialectic relationship between discourse and social structure or stratification.

Clearly, neither political nor social power is in itself a stable phenomenon but might undergo change together with the context. Groups struggle to obtain more positive constructions or might become more powerful by mobilising for political action (Ingram/Schneider 1997, 110), a process that can lead to the reconstruction of problematisations. Fairclough (1992, 95) maintains that an order of discourse and discursive practice can be seen as a discursive facet of hegemonic struggle, which contributes to the reproduction or transformation of the existing order of discourse. According to Fairclough's (ibid., 91-95) usage of the term, hegemony does not necessarily depend upon the existence of a coherent formation of society, but rather on the presence of a dominant complex of power-knowledge that can channel conflict debate and, by doing so, gives precedence to a particular resolution. Thus, the outcome of struggles heavily depends on which social forces succeed in gaining dominance over the discourse and channelling the debate in a particular direction (ibid., 96).

1.3 Discourse Stimulation

"Rather than a massive censorship, beginning with the verbal proprieties imposed by the Age of Reason, what was involved was a regulated and polymorphous incitement to discourse." Michel Foucault (1998, 34)

Despite the differences existing between the approaches to policy discourse discussed so far, they focus on claims made by social movements or powerless social actors, which the authors consider to be generally acceptable and legitimate, such as claims made by feminists (Bacchi 1990; Fraser 1989a), by techno-sceptical actors (Fischer 1999), or by environmentalists (Hajer 1993), or claims of action against poverty or inequality. In these aforementioned approaches, policy discourse is the result of the politicisation of these claims and is analysed from the viewpoint of how these claims are either marginalised or reframed in terms of dominant interpretations or problematisations. The authors seek to explain the relevant power-processes for sorting out particular problematisations. Power structures in these studies are thus analysed and manifest in the suppression or reframing of marginalised discourses in order to stabilise power (structures). On the other hand, discourses (of marginalised groups) appear as attempts to destabilise power. This is most obvious in Nancy Fraser's (1989a) discussion of needs discourses in the late capitalist welfare state. The model of social discourse she develops is based on the assumption that 'oppositional discourse' by powerless groups is the starting point of (policy) discourse, countered by what she calls 're-privatisation discourse' (ibid.) of powerful actors. Indeed, whilst there are also several reservations about viewing discourse as a means of destabilising power in these analyses, what remains underexposed are the discourses that are not initiated by marginalised powerless groups but by powerful social actors. At a more general level, the stimulation of policy discourse as an effect of power and as a means of stabilising and increasing power is not usually considered. Therefore, in the following, I suggest that in the analysis of policy discourse the discourses that are incited by powerful social actors be included, or, to put it more generally, the analysis should also consider the stimulation of discourse as a powerful activity directed at the stabilisation or expansion of power. Against the background of Michel Foucault's notion of the productivity of power, the approaches to policy discourse introduced above will be critically discussed with special emphasis on whether or not and how the dimension of discourse stimulation can be integrated in the analysis of policy discourse.

Discourse Stimulation as a Form of Power

As was pointed out in the previous chapters, problematisation is not simply an act of recognising a problematic situation, but is deeply embedded in social and discursive power configurations that affect processes of problem construction. Carol Bacchi (1999) and Nancy Fraser (1997b) question the presumption that problematisation and politicisation (including the initiation of discourse) can simply be viewed as recognition of a disturbing social condition and thus as a success in itself in terms of the desired outcomes. In her exploration of "the politics of need interpretation", Nancy Fraser (1989b) demonstrated that the successful politicisation[8] of an issue does not necessarily lead to the outcomes its initiators intended. Instead she maintained that it does not even guarantee the requirements that would make these outcomes possible. Fraser (1989a, 164) emphasises particularly

8 'Politicisation' here refers to the location of an unwanted social situation within the realm of political conflict or discourse, that is, to publicly question the acceptability of or consent to a certain situation and to, at best, induce a transformation.

that socially accepted forms of public discourse available for problematisation or problem discourse are not necessarily adequate and fair—in fact they might even work in favour of the interests of dominant social groups. Thus, a discourse following the successful problematisation by marginalised groups may still end in favour of dominant groups. Hence, Fraser describes political discourses as struggles over interpretations—or we could say problematisations—rather than over predetermined issues. Thereby she emphasises the fact that it is an essential feature of political conflicts that the interpretation of problematic situations, and not only their status within the political realm, is a political stake. According to this view, political discourse depends on the dominance of a (certain) problematisation to which participants are subjected. Therefore, in reference to Fraser's (ibid.) notion of the "politics of need interpretation", we can speak of *the politics of problematisation.* In this respect, Fraser (1997b) counters liberal theories pertaining to the democratic public, which tend to overlook the fact that forms of public discourse are constructed according to societal power structures and hierarchies and, correspondingly, reflect them. Such forms of public discourse include, for example, discourse boundaries, positioning, limiting the participants, or the division between public and private. Social power configurations not only have a significant impact on the discourse process and its outcome, they themselves are also highly contested within the discourse. Fraser contests that public discourse does not necessarily offer the possibility for power to be neutralised, asserting that it might, on the contrary, even offer the possibility to stabilise power, as distinct from a Habermasian understanding of an *ideal* discourse as a sphere of argumentation free from power and directed at finding a common consensus (Habermas 1995, 144). The politicisation of an issue and the realisation of a public discourse do not necessarily imply emancipation from power, Fraser maintains. Instead, established forms of discourse are intrinsically power-laden (Fraser 1989a; b; 1997b). Fraser concludes that political discourse is a "site of struggle where groups with unequal discursive and non-discursive resources compete to establish as hegemonic their respective interpretation of legitimate social needs" (Fraser 1989a, 166), and an intervention in political struggle can only be successful (if at all) if discursive and societal power configurations are addressed and challenged.

As a way of reducing power effects within discourse, Fraser (ibid.) demands that interpretations be justified and substantiated. A relevant factor is an inclusive and egalitarian process of generating interpretations. The best interpretations are obtained in a communicative process that is democratic, equal, and fair to the highest possible degree. The consequences of a widespread acceptance of an interpretation should be as disadvantageous as possible, challenging social patterns of discrimination and dominance (ibid., 182). After all, Fraser assumes that discourse can in principle be (largely) freed from power, whereby she refers to the Habermasian concept of the "public sphere" as an "institutionalised arena of discursive interaction" (Fraser 1997a, 70). Fraser states that, despite her critique of Habermas's ideal, "something like Habermas's idea of the public sphere is indispensable to critical social theory and to democratic political practice" (ibid., 70f). While this somewhat contradicts her critique of discourse elaborated upon above, Fraser points out the lack of an *alternative* to discourse in democratic societies, leading to the need to find forms of discourse that eliminate instead of simply bracketing out social inequalities (ibid., 92).

Altogether, Fraser scrutinizes the ways in which powerful actors react to discourses or problematisations of less powerful groups in order to conserve and defend their own power. She seeks to understand how powerful actors use their power to direct discourse in distinct ways, thereby authorising certain problematisations as well as influencing how social power structures support their dominance also in discourse. Fraser focuses on the question if and how less powerful social actors can make their case heard—assuming that they have a legitimate point—and suggests ways of how to reduce power within discourse. Power, in this view, appears as a means to suppress or reframe certain claims, problematisations, or discourses in order to stabilise power configurations within society. Discourse, on the other hand, appears as a weapon for marginalised groups or individuals to put power structures up for negotiation. Discourse and participation in discourse remain the only means—while certainly not a guarantee—for destabilising power.

However, especially in the field of reprogenetics and biomedicine, as well as in science policy more generally, problematisations (e.g. of 'too restrictive regulation') are frequently introduced by powerful actors, such as the state, members of government, policy officials, and scientific professionals or organisations etc., who constantly and increasingly demand and induce public debates on such issues (cf. also Irwin 2006; Irwin/Michael 2003). These discourses cannot simply be analysed as *reactions*—to use Fraser's term—to the discourses that are initiated by powerless or suppressed social groups. Nor can they be analysed solely in terms of reframing suppressed discourses or as attempts to stabilise the status quo. On the contrary, they are explicitly directed at transforming the status quo and promoting deregulation of science towards more permissiveness, as will be shown in the empirical section of this study. Therefore, the analysis must integrate the issue of how and if discourses initiated by powerful social actors promote expansion or exercise of power. Regarding this issue, Michel Foucault's notion of power being *positive productive* rather than negative suppressive affords insight especially into what he called the "incitement to discourse" (Foucault 1998, 17).

Michel Foucault explicitly dismissed an understanding of power in the juridical sense in favour of a conception of power in terms of positivity and productivity. In his renowned book, *The Will to Knowledge—The History of Sexuality I*, Foucault (1998, 17) addressed what he called a "discursive explosion" on sexuality as opposed to the assumption of mere repression of discourse and sexuality. Related to his analysis of bio-power, Foucault believed that "discourses on sex did not multiply apart from or against power, but in the very space and as a means of its exercise" (ibid., 32). While Foucault explicitly did not claim that prohibition, suppression, or exclusion did not exist, he rejected the view that power did not go beyond these aspects (ibid., 12).[9] Instead, he asserted that an understanding of power in juridical terms as a "mechanism destined to say no" (ibid., 12) serves as incitement to discourse and, by doing so, to exercise power and increase it. At the same time, the positive-productive qualities of power are concealed. In this respect, one of the essential questions in Foucault's work focused on the integration of subjects in a power nexus via their participation in discourse (Foucault 1994; 1998). As Foucault suggested, an important means of integrating subjects is the "repressive hypothesis", which understands power as opposed to truth and stimulates discourse as a form of opposition to power (Foucault 1998, 16ff). Power, in Foucault's terms, masks itself by producing an allegedly 'liberating' discourse, which is "really part of a larger deployment of modern power" (Dreyfus/Rabinow 1982, 130) and intensifies its efficiency. Discourses of liberation based on the presumed contradiction of truth and power operate as a means of stabilising and increasing power. Indeed, Foucault was interested in the question

9 In *The Order of Discourse* (1974), Foucault adhered to an understanding of the relation between power and discourse as based on restriction and negation, which he later explicitly abandoned (Foucault 1978, 105).

of how power manages to stimulate discourses rather than the issue of restricting it (Foucault 1998, 17 & passim).

From a Foucauldian perspective, the focus on the negative character of power—effectively suppressing or reframing particular (allegedly) legitimate discourses, claims, or problematisations—falls short of the positive productive character of power, especially in regard to the stimulation of discourse. Despite recognition of the positive productive effects of power in semantic terms, an analysis of discourse stimulation as an effect of the productivity of power remains underexposed, especially if we perceive the use of discourse by powerful actors simply as a means of suppressing other actors' discourses, as, for example, Nancy Fraser seems to suggest. This view bears the danger of misunderstanding discourses initiated by powerful actors simply as a democratic 'offer to openly negotiate certain (suppressed) claims' or as 'mere answers to demands for discourse by powerless groups'. At the same time, the assumption that the character of power in policy discourses is primarily suppressive insinuates that discourse—or certain forms of it—generally has a liberating function. On these grounds a demand for discourse is implied, even if it is restricted to a *certain* discourse. If successful, discourse appears as a process that is, in principle, directed at a reduction of power, which, in fact, was Foucault's main point of critique of the "repressive hypothesis" (ibid., 16ff). Discourse, in Foucault's terms, must be understood as a form of power production in two ways: On the one hand, through discourse objects are created (Foucault 1974, 37) in the sense of problematisation discussed above. On the other hand, discourse is a form of power acting on subjects by integrating them into the power nexus (Foucault 1994, 1998). Thus, with Foucault, we can analyse discourse stimulation by powerful social actors as power-laden attempts to challenge the status quo and also as a form of power production via the invitation to participate in discourse. There is, in fact, no reason to suggest that powerful social actors might not use discourse to transform the status quo for their own benefit. Of course, problematisation or politicisation entail the stimulation of policy discourse as well as the wish to transform the status quo, otherwise we could not reasonably speak of '*politic*isation'. However, if we do not comprehend discourse as a neutral tool for describing reality but as a powerful device for *making sense* of it, its stimulation must be analysed as an effect of power as well as a form of producing power. *Making sense* not only applies to the production of an object of discourse—the policy problem—but also to the delineation of the epistemic space within which discourse on it can proceed. Thus, problematisation of an issue not only creates its object—the 'problem'—but simultaneously suggests the necessity to talk about it whilst also delineating the politico-epistemic space for its discussion: Discourse not only creates meaning but likewise the space for experiencing and talking about the objects it itself produced, in addition to inviting subjects to participate.

I suggest that the implicit adherence to the negative and suppressive effects of power in discourse in many post-positivist policy analyses stems from a focus on particular kinds of political problems and discourses: Namely, on those initiated by powerless social actors or groups and understood to be *legitimately* addressed while being suppressed by powerful social actors, who wish to *re*-depoliticise problems and defend the status quo. Problematisations introduced by powerful social actors are usually not considered. Thus, these approaches are beneficial for comprehending how power works within discourse, that is, for comprehending the exclusion or reframing of discourses or problematisations of suppressed or powerless groups. These approaches fail to grasp problematisation processes and discourses initiated by powerful social actors or, indeed, discourses that are not directed at the emancipation from power but at its stabilization. I certainly do not deny the troubling or problematic character of many societal or individual conditions problematised by social movements or powerless actors nor the need for the analysis of the discursive processes or, least of all, the need to improve these conditions. However, critical social science must also scrutinize the fact that powerful social actors likewise problematise issues and initiate policy discourses in order to achieve certain ends, that is, to transform a social or political situation they find unacceptable. There is no reason whatsoever to assume that powerful actors should or would not problematise or politicise certain issues and start a discourse on them in order to change the status quo rather than merely defending it in reaction to oppositional discourses—even more so because they have much easier access to the necessary resources to do so than powerless groups or individuals have.

On these grounds, the analysis of the power of discourse stimulation has to be integrated into the study of policy discourses on reprogenetics. This means that the notion of *problematisation* has a twofold meaning: it not only refers to the semantics of problem construction but also to the inherent call to talk about and change the 'problem' as a means to stimulate policy discourses and to invite participation.

Before I go into the empirical analysis of reprogenetics policy discourses in Great Britain and Germany, I will introduce the ethnographic approaches studying the relationship between science, technology, and society that are more narrowly engaged with the investigation of the governance of *science* than post-positivist or interpretive policy analysis is. Science, technology, and society (ST&S) studies centre on the relationship between science and society, and problematise specifically a traditional, that is, expert-oriented, technocratic, and paternalistic model of science governance. As a result, scholars, such as Brian Wynne, Alan Irwin or Sheila Jasanoff, have (partly successfully) promoted a more integrative model that includes the public, more knowledge forms including also non-scientific forms of knowledge, as well as normative values within science policy. More recently these scholars have also been engaged in a critical evaluation of the success of their critique, that is, they have addressed new modes of science governance marked by an increasing participation of the public. In this respect, especially Alan Irwin (2006) has already hinted at the ambivalences and restrictions of public participation in (state-sponsored) science policy discourses, which is of great value for the understanding of what I call discourse stimulation.

2 The Relation between Science, Technology, and Society

2.1 From a Deficit Model to the Ethnographic Turn in Public Understanding of Science

Investigations of the public understanding of science (PUS) deal with the issue of how the public comprehends and evaluates scientific developments as well as the relation between scientific knowledge and the acceptance of science and technology (Durant/Joss 1995). The study of the public understanding of science began when strong conflicts in the public sphere were triggered by technological and scientific developments as well as when scepticism about the positive effects of techno-scientific advances within the public increased or grew more obvious (ibid.; Felt et al. 1995).[10]

10 Cf. Wynne (1996, 44ff), who claims that there was not a (sudden) shift from trust towards mistrust, but rather that irreflexivity of science within the public had been inadequately identified with trust, which he claims is an illusion (ibid., 47) that stems

The great conflicts arising over nuclear power and environmental issues were interpreted as the result of insufficient information and education on the risks of techno-scientific developments and practices within the public. The traditional PUS approach suggested that greater public acceptance of science and technology would grow in relation to an increase in science education (cf. Durant/Joss 1995; Royal Society 1985). In this situation, research on the public understanding of science aimed at greater acceptance within the public. A report by the British Royal Society (1985) suggested that everyone, and especially those in decision-making positions, needed to understand science to a certain degree and therefore recommended more scientific education. A basic assumption of the Royal Society was that national well-being depended on science and technology, necessitating science to be 'publicised' in order to make it acceptable (ibid.). The report strongly influenced the development of methods to estimate public opinion on and knowledge of science, technology, and science education programmes. Often using questionnaire methods, the traditional approach to PUS was directed at the quantification of what is called the 'scientific literacy' of the public (Irwin/Michael 2003, 19ff). Scholars such as Wynne (e.g. 1995), Irwin (e.g. 2003) and others of the ethnographic branch within ST&S have shown that especially the presupposed authority of 'scientific knowledge' implied in traditional PUS is inappropriate for understanding the relation between science and society. In fact, they criticised the understanding of the relation between science and society based on a belief in scientific truth, expert authority, elite decisions, and technical control within science policy (cf. Plough/Krimsky 1987; Fiorino 1989)

In a critical assessment of a science-based notion of knowledge within PUS, ST&S studies point out that policy-makers and policy experts as well as conventional PUS studies are ignorant of the context dependency of knowledge as well as knowledge acquisition and processing (Irwin/Michael 2003). PUS is predominantly criticised on account of knowledge being inherently stratified and the implication that one could and should differentiate between 'correct' and 'false' or 'better' and 'worse' understandings of scientific issues. According to ST&S scholars, this would lead to an inappropriate decontextualisation of knowledge and understanding, a devaluation of certain 'local' knowledges (cf. e.g. Carr/Levidow 1997; Irwin/Wynne 2003; Wynne 1992). This technocratic model of science policy

from a systematical suppression of alternative knowledges, and the marginalisation of ambivalences and dependencies within the public (ibid. passim).

would be unable to address unknowns as well as moral or scientific uncertainties and ambivalences (cf. Jasanoff 1994).

The understanding of the origin of public conflicts or unease about science in traditional approaches to PUS has been called the "deficit model of the public" (cf. Irwin/Michael 2003, 21ff; Wynne 1992; Wynne 1995), which one thought could be rectified by scientific information and education. ST&S scholars have not only shown that a deficit model of the public does not adequately describe the phenomenon of public scepticism towards techno-scientific developments, but also that this model is based on an elitist and ignorant understanding of knowledge. As Irwin and Michael (2003, 27) have pointed out, already the methods of traditional PUS, such as questionnaires presented to individuals, are embedded within a scientific framework that comprises a decontextualised and foreshortened understanding of knowledge centring on science. This framing would constrain the expression of responses and inevitably lead to the construction of an ignorant and uninformed public, resembling a self-fulfilling prophecy. At the level of public policy, a traditional view of knowledge and public understanding of science implies that "while science needs to be 'communicated' to public groups the rational basis for science policy should remain 'best science'" (ibid., 41), which effectively insulates policy-making from public concerns.

In contrast, an ethnographic approach to public understanding of science[11] stresses the validity of what is called "lay epistemology" (Irwin/Michael 2003, 28) and the intellectual capacities of non-scientific actors as well as the socio-cultural embeddedness of knowledge—including scientific knowledge. In his widely acknowledged study on policy-making in an agricultural area in Cumbria after the Chernobyl catastrophe, Wynne (1992; 1996) revealed that it was the ignorance of scientific experts and policy-makers towards forms of knowledge other than scientific, such as knowledge about local life or embedded agricultural issues etc., that reduced trust in expertise and science policy-making rather than public or local community ignorance (in this case Cumbrian sheepfarmers) of scientific knowledge. Wynne illustrated that inconsistent policy decisions and an obscure institutional landscape as well as an alleged but, in terms of the local knowledge, unrealistic certainty of expertise severely undermined trust in policy-makers and experts. Not only was the sheepfarmers' non-

11 Irwin and Michael (2003, 28) speak of the ethnographic turn in PUS.

scientific knowledge disregarded, but even what they experienced was contradicted by the scientists' proclamations. Scientific laboratory knowledge was simply transferred to the Cumbrian fells and thus inappropriately over-extended into a very different context. The self-assurance of scientific experts, however, hindered them from learning from the local 'experts'. Local knowledge was severely devaluated, leaving the Cumbrian sheep-farmers with the feeling that they were completely dependent on expert decisions—which in fact they were—and alienated from their own local knowledges. At the same time, powerlessness and dependency on expert systems was concealed by their behaviour "as if" they trusted them (Wynne 1996, 52).

Especially in science policy, Wynne detects the institutionalised socio-cultural stratification of knowledge, which lends privileged authority to particular epistemological cultures—science and expertise—and marginalises others. Wynne (ibid., 46) states that, through boundary work, knowledges other than those based on science are systematically suppressed and concealed behind a rhetoric of objectivity and rationality (ibid., 74). Through 'boundary work' a demarcation between science and varieties of 'non-science' is constructed in flexible and historically changing ways, and a "public image for science by contrasting it favourably to non-scientific intellectual or technical activities" (Gieryn 1983, 781) is created. A common device to marginalise public knowledge of scientific developments is to separate the 'objective' and supposedly 'neutral' from the 'emotional' and 'subjective' as disconnected dimensions, declaring the 'emotional' to be deficient. Due to this separation the 'objective' and supposedly 'neutral' scientific dimension remains unreflected (Carr/Levidow 1997; Wynne 2001). In opposition to the objective-emotional divide, ethnographically oriented studies on the relation between science and society set an understanding of knowledge as epistemic culture, in which scientific as well as other forms of knowledge are *all* understood to be culturally embedded, intellectually sophisticated, and legitimate. Scientific knowledge, in this view, is not intellectually superior or more valid than other—for example local—types of knowledge.

2.2 Towards more Participation and Integration of the Public in Science Policy

"Equally […] at the heart of the 'new' reside some very 'old' assumptions." Alan Irwin (2006, 304)

As has been already indicated, it is apparent that traditional PUS as well as ethnographic ST&S scholars address problems of public response to science and technology development or—more generally speaking—a conflict-ridden and problematic relationship between science and society. Especially research done by the ethnographically oriented scholars is not only concerned with confronting the academic framing of the deficit model but also with the ensuing implications for the policy process and policy-making. They recommend that their research findings be acknowledged and integrated in practical politics (Irwin 1995; Irwin/Michael 2003, 41ff). Most importantly, these scholars have promoted the participation of the public or concerned actors as well as the integration of different forms of knowledge in the governance of science and technology. Thereby the latter is not to be comprehended as an alternative to scientific knowledge but serves as inspiration for finding collective self-conceptions and thus arrive at better policy decisions (cf. e.g. Wynne 1996, 78). In a following second step, ethnographic ST&S scholars analysed if and how the awareness of the problematic relationship between science and society as well as their critique of a deficit model influenced or transformed practical policy-making (Irwin/Michael 2003, 40).

In fact, Irwin and Michael (ibid., 51f) and Irwin (2001) state that British policy-makers had obviously learned from deficit as well as ethnographic approaches to PUS and that they found evidence for a "changing social climate for science-public relations" (Irwin/Michael 2003, 52). Policy-makers themselves tend to speak of a 'need for dialogue between science and the public' instead of a scientific monologue: they demand 'transparency' and 'openness', refer to 'public values', and support the 'necessity to clearly lay open the boundaries of knowledge' as the prerequisites of building public trust.[12] While the deficit model assumes that the concern about or critique of scientific development are the result of a deficit in knowledge and information, which can be relieved via education and information,

12 This is also obvious in the Royal Society's programmes. Cf. webpage of the Royal Society at: http://royalsoc.ac.uk, Mar 10, 2007.

today we can observe the emergence of more integrative modes of governance, such as public consultations or discourses, suggesting a departure from a deficit model also within science policy itself (Irwin 2006). In addition to the scholars themselves, who are now members of advisory boards, ST&S studies' critique and language seem to have been successfully integrated into science policy, leading to a "putative" transformation from "technocracy to democracy" (ibid., 301). Also, at the level of public policy more generally, Hajer and Wagenaar (2003, 3) suggest that hierarchical science-based and state-centred policy-making has lost its heuristic force, and now new ways of finding legitimate solutions to societal problems are sought. They assert, therefore, that not only societal problems be on the agenda but that also politics and policy-making are in fact problematised. Different new forms of dialogue as well as the establishment of committees and councils "all suggest that politics is not simply concerned with outcomes", and "we quarrel about the rules of the games themselves" (ibid., 3). Many scholars (such as Braun/Herrmann 2000; Carr/Levidow 1997; Fairclough 1992; Hajer/Wagenaar 2003; Irwin 2006) perceive an increase in discursive and participatory procedures at the level of political practices. This increase in discourse marks a change in political and social practice in which discursive participatory practices become more salient (Fairclough 1992, 8). In sum, we can observe significant transformations in science and especially in biomedical science policy-making: beginning with a technocratic, science-oriented model and then advancing towards more participatory, more open, and discourse-oriented forms of science governance (cf. also Braun et al. 2008). It is not only—or even predominantly—suppressed or marginalised groups who demand and support participatory discourses, but, in fact, powerful social actors, such as governments or scientific professionals.

However, despite the fact that the deficit model of the public—also within policy processes and policy-making—seems to be on the decline, Wynne (2002), Irwin (2006), and Kerr (2003) illustrate the fact that the notions of the supremacy of scientific epistemology and the need to educate the public are still very much alive. Despite a shift towards increased integration of the public voice and a trend towards public discourse in the policy process, they observe the old policy imperative of 'sound science'. "The modernist belief in science-led progress [...] is not being rejected [...], but is instead augmented with an assertion of the essential compatibility of (particular forms of) science and democracy" (Irwin 2006, 301).

Irwin (2006) and Wynne (2001; 2002) maintain that the assumption of the supremacy of scientific knowledge and expertise persists in framing processes, affecting opportunities for engagement as well as possible outcomes. "Control over the framework for engagement […] constitutes an important source of power" (Irwin 2006, 316). The agendas tend to be pre-framed in that they "assume the inevitability of technological development and commercial benefit from the development" (Kerr 2003, 216; cf. also Irwin 2001). In 'new' participatory modes in the government of science, ST&S scholars also criticise that the dominance of scientific epistemology successfully prevents self-reflective processes taking place in science (Carr/Levidow 1997). Thus, developments towards more participation and reflection "are severely compromised by the extent to which that dominant culture reinvents and extends its unreflexive founding commitments in the face of such critiques and public disaffection" (Wynne 2002, 472).

Irwin's findings on the British *GM Nation?* public consultation[13] suggest that scientism prevails, while more inclusive governing procedures are endorsed also by policy-makers themselves. It seems that "the new scientific governance advocates *both* increased openness *and* more professional/centralized control over risk-management. Although not necessarily in contradiction, the clear implication is that the possibilities for public challenge to the dominant institutional framework of risk management will be constrained" (Irwin 2006, 307, original emphasis).

As the persisting dominance of scientific epistemology and the rather strict separation of scientific and other forms of knowledge seem to offer little opportunity for public engagement (Irwin 2006, 306f), expanding discourses and participation opportunities in the domain of science policy must be critically assessed: Regarding the prevailing intellectual authority of science, it seems rather inappropriate to speak of a clear *shift* in policy. The impact of 'sound science' dominating in framing or problematisation processes has become apparent despite an alleged 'democratisation' of modes of scientific government.

This, however, is only one part of the story. As Irwin (ibid.) points out: apart from the persistence of old scientistic assumptions, new aspects related to the governance of science in fact now exist that go beyond a simple belief in the ability of a scientific elite to successfully govern science. In

13 *GM Nation?* was one the most complex public consultations in the field of science and technology policy in Britain, dealing with issues concerning genetically manipulated organisms (Irwin 2006, 309ff; cf. also Gaskell et al. 2003).

Irwin's terms, "a politics of talk" (ibid.) has emerged that is marked by an emphasis on openness, transparency, and public participation, undeniably calling for a critical analysis.

Particularly in the issue area of science, several authors detect new forms of debate directed at more inclusion and participation of the public, such as public consultation, dialogue, participatory deliberation, and events etc., as opposed to merely scientific debates restricted to experts. The new forms of debate 'acknowledge' public (sometimes 'ethical') concerns about techno-scientific developments (Carr/Levidow 1997; Irwin 2006; Kerr 2003; Wynne 2002). Regarding this phenomenon, Alan Irwin (2006) observes the rise of a 'new' deficit model: that is, a *deficit of public trust* in the ability of politics to successfully govern science. This deficit was regarded as the cause for raising concern about techno-scientific developments. The absence of trust takes the place of an absence of knowledge in raising concern for science policy-making (cf. also Wynne 2001). "Perhaps unsurprisingly, the central issue appears to be public trust in current mechanisms of science policy rather than a more fundamental reappraisal of the relationship between science and social change" (Irwin 2006, 306).

As suggested, the 'new' modes of scientific governance are the answer to the assumed deficit of trust. The transparency and openness of the public engagement exercises are comprehended as a means of convincing sceptical members to trust decision-makers in the policy process, a trust that had diminished especially after the BSE debacle (cf. Forbes 2004). Policy-makers detected a "crisis of confidence" (Irwin 2006, 307) in science leading to the suggestion that the relation between science and the public needs to be reorganised (ibid., 306). A remedy was sought for in the openness, transparency, and dialogue with the public (ibid. 308). However, despite enthusiasm towards participation, Irwin detects that new inclusive forms usually offer little or no scope for public engagement in the *development* of science. One sign of this is the strict separation between scientific evidence and ethical or social considerations in new participatory forms of public engagement (ibid., 305f). Carr and Levidow (1997) come to similar conclusions in their analysis of public debates over agricultural biotechnology, which are marked by supplementing 'risk' discourses with 'ethical' ones. While the integration of 'ethics' in state sponsored debates was meant to be a response to and to integrate public concerns over biotechnology, the authors suggest that it rather served to protect science against critical ('ethical') reflection: Rather than being considered an integral quality of 'risk' discourse, 'ethics' was introduced as an issue separate from 'risk', thereby implying that 'risk' was a value-free scientific category. According to Carr and Levidow, while 'risk' is itself the result of value judgements, that is, what is called the "implicit ethics" of 'risk' (ibid., 29), value questions were located in an 'ethics' discourse distinct from that of 'risk'. 'Risk' is thus framed as 'objective science' not subject to critical reflection. "This form of ethics serves to compensate for the unacknowledged value-choices and institutional commitments already embedded in R&D priorities" (ibid., 29). Thus, integrating 'ethics' served to stabilise science-based 'risk' discourse instead of reflecting or scrutinising it: Sound science was not replaced or even reflected upon in the new modes of scientific governance; indeed, public concerns were framed in the terms of the 'hidden agenda', which is still "sound science" (Irwin 2006). This prevented 'sound science' from being scrutinised as well as, more generally, avoiding the questioning of the character and direction of scientific development.

Also in the context of debates on genetically modified organisms (GMO), Wynne (2001) analysed the growth of institutionalized 'ethical' debate, which appeared together with 'recognition' of public ethical concerns and a rise in participatory, dialogic initiatives as opposed to mere 'risk' debates. In Wynne's analysis, constructions of the public, its capacities, and forms of judgements imposed through scientific policy culture are a major concern. He attempts to comprehend the meaning and effects of "dominant characterizations of public concerns about GMOs, especially in the relatively new domain of ethical concerns that form a recent and growing agenda through which public responses are recognized, constructed, and expressed" (ibid., 446). Wynne asserts that public concerns are "apparently given greater autonomous policy weight" (ibid., 446) when defined as 'ethical concerns', because they do not appear to be subordinate to scientific knowledge, as is the case with 'risk' concerns. However, Wynne detects that, at the same time, dominant definitions of 'public ethical concerns' systematically deny and erase central aspects of typical public judgements of science policy (ibid., 447). He points out that 'ethical' concerns in biotechnology are commonly defined either scientifically as weighing up the consequences, or as pure "matters of private and individual choice which can be resolved by market mechanisms alone" (ibid., 446; cf. also Carr/Levidow 1997, 38). Public concerns are framed as arising from 'emotions' and 'intellectually vacuous' ethical concerns, rather than as "what they are, which is public judgements of the quality of existing knowledge, and of the exaggerated claims made for it by scientist and the policy bodies they advise" (Wynne 2001, 457; cf. also Wynne 1996).

All these studies suggest that 'ethics' discourses on agricultural biotechnology or genetics did not, and could not, serve as counter-discourses to scientific ones, but, on the contrary, that they strengthened science's force (Carr/Levidow 1997). Talking ethics as an issue separate from science ultimately freed science from scrutiny. By providing an outlet for public concerns, the 'ethical' agenda allowed the 'objective'-'subjective' divide to be intensified further (Wynne 2001, 445). In this respect, the integration of 'public ethical concerns' and the shift to more participatory modes of science governance served most obviously to stabilise rather than reduce power. Thus, to simply speak of *new* forms of governing science is not only misleading, but also obscures the fact that the 'old' model of 'best expertise' is maintained in science policy. Despite a general agreement on the need to integrate the public in science policy, there is, at the same time, a shared assumption that public participation does not necessarily facilitate a broad change in science policy in terms of a 'true' retreat of science in the policy process. Nevertheless, the authors holding this view do not simply dismiss new participatory modes of science policy as rhetorical fake initiatives, but instead they want to empirically understand how participation and scientism work together and with what effect (Irwin 2006, 316): That is, they analyse the new public debates also in terms of what they *do* rather than merely what they do not do, or, to use the words of Michel Foucault, they analyse the debates not only in terms of their restrictive but also in terms of their productive qualities.

Creating the 'Confident' and 'Responsible' Consumer

While the "historical commitment to progress through science is maintained" (Irwin 2006, 308), "public anxieties [...] have become a recognized feature of science and technology policy" (ibid., 310), so that referring to the need for re-establishing public trust became essential in statements in science policy. Irwin points out that, in addition, economic pressures on science-based innovation are very powerful, leading to the search for an approach to public engagement that "will permit rather than impede scientific and technological development" (Irwin 2001, 3). The goal of public participation, of a "politics of talk" according to Irwin (2006, 303), is to generate consensus on or acceptance of science and scientific development in the public sphere. In this case, however, it is achieved through engagement. The target is to generate the "confident consumer" (Irwin/Michael 2003, 52) in a science-led society. When perusing, for example, the Royal Society webpage, it is apparent that, although "it debates scientific issues with the public",[14] it is nonetheless devoted to scientific progress: "The Royal Society, the national academy of science of the UK and the Commonwealth, is at the cutting edge of scientific progress".[15] "A dialogue with society", as the Royal Society states in its Science and Society Report of 2004, "is important for science to have a licence to practice" (Royal Society 2004, 7). Openness, transparency, and public engagement should thus not limit scientific progress but "create a more open and reflective culture where new scientific possibilities can be fully realized. [...] the historical commitment to progress through science is maintained: the challenge is to find more inclusive methods to achieve such progress" (Irwin 2006, 308). Boosting participative dialogue is intended to secure "science's licence to practice" and not restrict it (ibid.). Thus, the question arises: "What scope can there be for dialogue when the direction is already set?" (ibid., 316).

An important feature of *GM Nation?* was a particular concern to avoid domination or subversion of the debate by the "special interests" of interest groups (ibid., 311). Despite an ongoing public conflict over genetically modified organisms, *GM Nation?* addressed particularly 'uninvolved' and 'uninterested' individuals; special measures were introduced to activate hitherto inactive individuals (ibid., 311). In line with Wynne's assertion about including 'ethical' concerns in GMO debates by consistently framing them as '*individual* matters' (Wynne 2001, 458) (and thus not matters of interests groups or political activists), Irwin's observations suggest that within public debates the 'general public' is constructed as uninterested and innocent (Irwin 2006, 311), implying that interest groups are not the general public and that interests are not and should not be part of the debate. "The implicit premise here—as elsewhere in this discussion—is that organized groups represent a problem for this form of public debate and conversely, that 'true' public opinion must be held by those without 'fixed views'" (ibid., 312).[16] As Irwin states, in this context organizers were keen

14 Royal Society webpage at: http://royalsoc.ac.uk, Mar 10, 2007.

15 Ibid.

16 Despite the fact that *GM Nation?* had addressed particularly individuals, it was accused of not having engaged with the wider public but instead with 'interest groups' accom-

to not have a yes-or-no answer to GM crops, but *GM Nation?* should instead "establish the nature and full spectrum of the public's views" (ibid., 312). The result of coupling of a new mode of governance with the construction of an innocent public, as Irwin states, is a departure from a political culture "where previously the treatment of controversial technologies has been adversarial in character (as in media treatments and public protests)" (ibid., 315). Techno-scientific issues, we should add, are not treated as issues of adversarial interests of different actors—that is, as political issues of conflict and sometimes combat—but as various (non-political) views of an unrelated aggregation of individuals who need to be integrated in debate. Anne Kerr goes even further, suggesting that, despite a rhetoric of public engagement and 'ethical' concerns, a *right* to public participation might sometimes "slip into the suggestion that they have a *responsibility* to participate" (Kerr 2003, 217, emphasis added) and a responsibility to educate themselves and engage in 'informed' debate.

If we look at the findings of Irwin, Wynne, Carr/Levidow, or Kerr concerning the proliferation of public debates on biotechnology and the increased integration of uninterested parties, we clearly do not find mere incentives for 'reprivatisation' discourses (Fraser 1989a) nor the intention to exclude participants intended to sustain the status quo, but we certainly find *more and more* participants in *more and more* discourses initiated by powerful actors, such as the state, professional organisations, or advisory boards. Thus, we can speak of 'discourse stimulation' in the sense discussed in the previous chapter. Discourses, such as *GM Nation?*, are initiated in order to instigate participation in the discourse especially among hitherto uninvolved individuals (Irwin 2006, 311) so as to create confident consumers and an environment in which "science's licence to practice" is secured and not restricted.

To sum up, the ethnographic approach within ST&S studies provides us with the necessary vigilance also towards modes or processes directed at the governance of (bio-)scientific developments that integrate critique of and work against technocracy and expertocracy. For the analysis of (new) public 'ethical' discourses in the area of biomedicine and reprogenetics, which are meant to respond particularly to dangers and risks of new biomedical and reprogenetic developments and keep them in check, the findings of ST&S studies are of great value, especially with regard to the issue

---

panied by scientists, who warned against a hostile social climate towards British science that could 'demoralize' scientists (Irwin 2006, 311).

of whether or not 'ethics' in actual fact forms a counter-discourse to scientific discourse. The studies discussed above have revealed the ambivalences of (state-sponsored) public discourses and show that a commitment to public engagement, to the integration of (ethical) concerns and critique, to debate, transparency, and openness does not exclude a similar commitment to scientific and economic progress. Indeed, the latter has not only survived public debates but has even been strengthened by them.

Policy Debates on Reprogenetics in Great Britain and Germany

The German and British policy discourses on reprogenetics gained momentum in the wake of the announcement of the birth of the cloned sheep Dolly in 1997 (GB) and the wish to perform pre-implantation genetic diagnosis (PGD) in a particular case in 1995 (G). Whilst embryo research, reproductive and other related technologies and practices were on the political agenda already prior to this, Dolly's birth and connected or assumed scientific possibilities triggered claims for the revision of existing legislation and public policy in the UK. In Germany, the demand for permission to carry out PGD, a practice not regulated by law but considered by many to be prohibited, was in itself interpreted as a claim for revision. The problem faced by policy communities was apparently—at least initially— whether or not the existing legislation, the British *Human Fertilization and Embryology Act* 1990 (HFE Act), covered the application of the Dolly technique to human beings, and whether or not PGD was prohibited by the German *Embryo Protection Act* 1990 (*Embryonenschutzgesetz*, EschG). Different actors engaged in and triggered off policy debates about the de- or re-regulation of these practices. However, the actors did not agree on what they considered was the problem that necessitated a new policy. A closer look at the debates reveals a striking disagreement on the level of problematisations. Reprogenetics was a "wicked problem" (Fischer 2000), which was controversial not only on the level of finding solutions but already on the very level of defining the problem itself. What is more, problematisations noticeably changed throughout the years: the 'problem' was constructed and reconstructed within the policy discourses themselves, and the problems that 'were solved', in the end, differed significantly from the initial ones. Although the processes of (re-)constructing the problem differed in both countries, the debates in Germany and Great Britain can, in different ways and to varying degrees, both be described as *struggles about problematisations.*

In the following I will present the different phases of debates with regard to the respective problem interpretations present at a particular time. In particular, the focus lies on how problematisations changed throughout the debate, how these changes came about, and who were the key actors initiating these changes. Especially the question of discourse stimulation will be emphasized, that is, the question of who initiates discourse through problematisation, how certain problematisations triggered discourses, and with what effects.[17] I will not discuss the (moral) issues raised in the policy discourses but recapitulate the process of construction and reconstruction of 'the problem'. The normative demand, which forms the background for analysis, stipulates that techno-scientific developments be open to political and societal shaping and control. For this reason it is essential that policy discourses provide the opportunity to challenge scientific developments and also the possibility to restrict or stop them. The political impact of problematisation and discourse stimulation on politicising scientific developments will be the focus of attention of this study.

Clearly, 'new' discourses or public conflicts do not simply arise out of the blue, but are embedded in and (positively or negatively) refer back to a historical, social, and discursive context. Thus, while an account of this context can never be complete, each country study begins with the introduction of the processes leading to the implementation of their respective regulations that came under scrutiny and which were finally changed in 2000 (GB) or 2002 (G), and includes the details of a few legal instruments, debates, and conflicts that seemed to have the most relevance for the policy debates under study as far as framing and problematisation processes are concerned.

3 Great Britain: To Clone or Not to Clone?

3.1 Historical Background

In the following, I will sum up the historical background of the debates on 'therapeutic' cloning in Great Britain. First of all, I will briefly outline the

17 I did not attempt to show which actors or discourses or whether or not they were *correct* or *incorrect* in their ways to interpret and present 'the problem'. On the contrary, the assumption of the *constructedness* of problems curtails a classification of problematisations following these categories.

conflicts subsequent to the birth of the first test tube baby in 1978, which led to the implementation of the Human Fertilisation and Embryology Act 1990 as well the contents of the act itself. This will be followed by a concise account of the conflicts on genetically modified organisms (GMO) and BSE, which led to a 'crisis in government' in Great Britain or were interpreted as such. Conflicts over GMO and BSE not only resulted in a decrease in public trust or were interpreted as having this effect but also brought about a transformation in policy-making. As Irwin (2006) and others have pointed out and as had been elaborated upon in the previous chapter, it was in this context that new participatory modes of policy-making increased, which likewise affected the ways to debate on or deal with reprogenetics.

3.2 The Human Fertilisation and Embryology Act

A Brief History

The British 1990 Human Fertilization and Embryology Act (HFE Act 1990) is the result of public and parliamentary debates following the birth of the first baby born after in-vitro fertilisation in 1978, which triggered much concern and unease in the British public. A few years later, in 1982, the Department of Health appointed the Committee of Inquiry into Human Fertilization and Embryology. It was referred to as the Warnock Committee and chaired by the philosopher Mary Warnock. Its objective was to consider the social, ethical, and legal implications of the newest and potential developments in assisted reproduction. The report (Warnock 1984) was published in 1984 and debated in both Houses. Although the Warnock Committee discussed a wide range of topics, such as in-vitro fertilization (IVF), the storage of human ova, or surrogacy, public as well as parliamentary debates focused predominantly on embryo research. Especially the fact that there was no law in place to regulate embryo research or to make scientists accountable for their activities in that field was found disturbing (Mulkay 1997, 3). Thus, the debates focused, to the greater extent, on the development and implementation of an act to regulate the issue. The Warnock report recommended, among other things, that a statutory authority to oversee and regulate research be established and provisions be made for infertility treatment such as IVF and artificial insemination (ibid., 16). With regard to embryo research and manipulation, the report suggested that it was considered a criminal offence unless it had been licensed by the authority. Also, the Warnock report suggested a 14-day limit for permitting embryo research, which restricted the use of embryos for research to no longer than 14 days after fertilisation.

According to Mulkay (ibid.; cf. also Mulkay 1993; 1994; Franklin 1993), the public debates following the publication of the Warnock report were extensive and rather polarized. There were two opposing camps on this issue (Mulkay 1994, 612). In 1985, conservative and pro-life views were quite prevalent in the UK. Enoch Powel, a conservative MP, introduced the Unborn Children Protection Bill, which aimed at prohibiting embryo research. He was supported by the anti-abortion group LIFE, which gathered over two million signatures from the British public. Pro-research groups were, as yet, not well organised, and a vast majority of conservatives as well as a slim majority of labour MPs opposed the research. Therefore, at the time, it was very likely that embryo research be prohibited. The Thatcher government delayed the second reading of the bill, which gave the pro-research lobby time to organize itself (Richardt 2003, 94). In the same year, the Medical Research Council, together with the Royal Society of Obstetricians and Gynaecologists, established a Voluntary Licensing Authority in order to document the ability of the medical profession to regulate embryological research and application within their own self-regulatory system. The pro-research lobby organisation PROGRESS was founded a year later, an umbrella organisation incorporating different pro-choice as well as pro-research groups (Mulkay 1997, 58f). Although Powell's private members' bill was rejected in the second reading, it was still doubtful if there was a majority in parliament to vote for embryo research. However, the result of the activities of the pro-research lobby was that public as well as parliamentary opinion "moved steadily towards support for embryo research" (Mulkay 1994, 613). In 1987 the government finally published the white paper, which allowed embryo research. This move outraged anti-research groups. However, campaigns against the research were overruled by campaigns of the pro-research lobby, especially PROGRESS, which had, in the meantime, grown. One of the most important achievements of the pro-research lobby, as Richardt (2003, 95f) writes, was setting abortion apart from embryo research by introducing the notion of the "pre-embryo". The coining of this term during the debates on the HFE Bill was "a political act" (Spallone 1989, 55f), which helped create an opposition between the legitimate research on the 'pre-embryo' and the illegitimate research on the 'embryo proper'. Mulkay (1994) makes it clear that the introduction of this term supported a change of opinion in parliament, which eventually accepted embryo research after initially rejecting it. A few weeks before the decisive parliamentary debate, the HFE Bill was amended to include a clause on abortion, namely to shorten the time limit for its legality from 28 to 24 weeks (ibid., 40). This helped the conservative government to express their pro-life commitment despite the fact that they supported embryo research (Richardt 2003, 96). What is more, the government introduced the HFE Bill in the House of Lords first, which is traditionally more 'research friendly' than the Lower House. The bill was finally passed in parliament on 23rd April 1990, that is, twelve years after the birth of the first IVF baby.[18] Apart from acknowledging embryo research as a legitimate area of science and passing legislation on it, another important outcome of the debates and controversies about embryo research was that

> "embryo research also emerged as a special case, that is, as an area of experimental investigation which was generally regarded as permissible and as clinically valuable, but which was deemed to be sufficiently dangerous to require careful regulation from outside in order to ensure that it did not develop in ways inimical to the interests of wider society" (Mulkay 1997, 2).

The Act

The 1990 HFE Act is very much in keeping with the recommendations of the Warnock report of 1984. It regulates, among other things, infertility treatment, the status of children born after IVF, cryopreservation of embryos, and embryo research. Schedule 2 of the act specifies activities for which a licence may be granted. It allows research on in-vitro embryos for up to 14 days after fertilization, if the goals of research cannot be met in other ways and if such research serves one of five research purposes: (1) to advance IVF; (2) to improve knowledge about congenital diseases; (3) to increase knowledge about the causes of miscarriage; (4) to improve contraception techniques; as well as (5) to develop methods of detecting genetic or chromosomal deviations in embryos before implantation (cf. Morgan/Lee 1991). The act also rules that consent to all activities regulated in the

18 For a more detailed discussion of the debates preceding the implementation of the 1990 HFE Act, cf. Mulkay (1993; 1994; 1997) and Franklin (Franklin 1993). For a comparison of the British parliamentary debates in 1990 with those in 2000/2001, cf. Parry (2003).

act must be provided by the person 'donating' the eggs, sperm, or embryos (HFE Act 1990, schedule 3). Furthermore, research must be licensed by the Human Fertilisation and Embryology Authority (HFEA), which was instituted under the act. Conducting research without a licence invites criminal sanction. The licensing of embryo research works on a case-by-case basis. While the UK has one of the most comprehensive legislations in this field, it also "legalizes a high degree of experimental freedom in relation to embryos" (Franklin 1999b, 71). The British regulation of embryo research links a high degree of permissiveness with an unusually high level of legislation, most of which refers to the composition of the HFEA and the processes and conditions of licensing.

The largest part of the HFE Act regulates the composition, function, and procedures of the HFEA, which was set up on recommendation by the Warnock Committee (Warnock 1984). The HFEA took up its office in August 1991, subsequent to the passing of the act. Apart from giving advice and reporting to the government on developments in infertility treatment and embryo research, the function of the HFEA is to license embryo research and IVF treatment and to control adherence to the HFE Act regulations. As an autonomous non-departmental government body, it regulates and inspects all clinics in the UK offering IVF, donor insemination, or egg, sperm or embryo storage as well as research centres that undertake embryo research. Its tasks are to provide advice and information to patients, donors, and clinics and to produce guidelines for clinics on licensed activities. Additionally, it has to keep relevant developments in treatment and research under review and likewise advise the Secretary of State.[19] The HFEA produces a code of practice as the guidelines for reviewing research applications. It is revised annually to accommodate technological and scientific developments.

In its first Code of Practice of 1991[20] (which, according to section 26, had to be approved by the Secretary of State and laid before parliament), the HFEA specified its guiding frame: Besides the requirements of the HFE Act, the HFEA is also guided by "the respect which is due to human life in all stages in its development", the "rights" of the people who are infertile, the concern for the welfare of the children, as well as "a recognition of the benefits, both to individuals and to society which can flow from

---

19 Cf. HFEA, About us at: http://www.hfea.gov.uk/AboutHFEA, Oct 29, 2003.

20 Cf. http://www.hfea.gov.uk/docs/ 1st_Edition_Code_of_Practice_1991.pdf, July 18, 2008.

the responsible pursuit of medical and scientific knowledge."[21] The code stated that all these guiding themes occasionally may stand in conflict with each other. However, the HFEA sought to reconcile them in a practicable way and in accordance with the HFE Act. The HFEA stated that it was their "aim […] to support the best clinical and scientific practice, while guarding against the undoubted risk of exploitation of people at a time when they may be particularly vulnerable."[22] Additionally, the HFEA stated that it wanted to keep the code under review in the light of experience and in order to be able to provide "up-to-date guidance"[23] at all times. The code has been constantly revised ever since; currently the seventh version is in effect.[24]

3.3 The Conflicts on GMO and BSE

Several events in the area of science and technology in Great Britain led to a climate of growing scepticism towards the ability of science and politics to effectively control scientific developments, such as the BSE crisis (cf. Forbes 2004) or public conflicts on genetically modified organisms (GMO). Both were interpreted also as crises in policy-making, meaning that there was scepticism towards whether or not politics were able to deal with risky issues and prevent unwanted consequences, such as infections (Creutzfeld-Jakob disease) or health or environmental risks related to genetically manipulated foods and crops (cf. Gaskell et al. 2003). The conflict about GMO began in 1996 when the UK imported the first genetically modified soy beans. In the 1990s, public support for GM crops and food in the UK decreased at an exponential rate. Especially in connection with other health scares, such as BSE, "[m]ore people reckoned that government and industry were doing a bad job, rather than a good job over biotechnology. Newspaper coverage reached a peak in 1999, and the average article became critical in tone" (Gaskell et al. 2003). Especially the BSE crisis was interpreted to be a sign of the limitations of expertise, revealing

21 Cf. http://www.hfea.gov.uk/docs/1st_Edition_Code_of_Practice_1991.pdf, July 18, 2008.

22 Ibid.

23 Ibid.

24 All versions can be found at the HFEA webpage, at: http://www.hfea.gov.uk/en/1538.html, July 18, 2008.

flaws in the British policy-making system (Forbes 2004, 343). Both BSE and GMO undermined public trust in scientific and regulatory systems, leading also, on the part of policy-makers, to the realisation that a 'deficit model' of the public was an inappropriate or unsuccessful conception for dealing with public mistrust (Wynne 2001). The British public rejected GMO, but, indeed, suspicion toward the regulatory process of biotechnology exceeded suspicion toward the technologies themselves (Gaskell et al. 1998; Gaskell et al. 2000).

Gaskell et al. (2003) conclude that while public attitudes towards what is referred to as 'red' biotechnology was usually much less controversial an issue in the UK than 'green' biotechnology, some authors presume that the government's hesitation to quickly make a law allowing research cloning was due to fears that scepticism towards GMO might spill over to issues of reprogenetics (e.g. Sexton 2000). To a certain extent, references in government responses to 'a need to reassure public trust' support this assumption, an issue which will be discussed in more detail below. The GMO as well as the BSE crises brought the notion of 'public trust/mistrust' and the 'need for public participation' into the focus of UK policy-makers (Irwin/Michael 2003; cf. Forbes 2004; Grove-White 2006). Public and media opposition especially toward GMO intensified, while the economic potential was increasingly advertised. Grove-White states that the British government, although it did not understand why GMO in particular had created such opposition and conflicts, was in the position of having to prove that different viewpoints were visibly integrated in policy processes on GMO. To restore public credibility was the most important goal pursued by the government (Grove-White 2006, 171f).

As had been pointed at in chapter 2.2, it was in this context that there was an increase in inclusive, participatory procedures for integrating varying demands in science policy-making and, most importantly, restoring public trust. It was within the GMO conflicts that public scepticism or critique of scientific developments was redefined—from being interpreted as arising from a knowledge deficit, which could be dealt with by education, to a trust deficit calling for participatory measures to involve the public. In the face of scientific developments, presenting public scepticism and concerns as an issue of 'trust' persisted.

3.4 Debate on 'Therapeutic' Cloning: How it all Began

The HFE Act and the issue of embryo research remained relatively uncontested until the announcement of the birth of the cloned sheep Dolly in February 1997 in The Observer (Feb 23, 1997), which caused an outbreak of public media debate in Great Britain on genetic manipulation of animals and, to a much greater extent, about the acceptability of human cloning.[25] Embryo research, especially the discovery and, later, destruction of thousands of frozen embryos in storage in 1996 because they could not legally be kept for more than five years, led to unease within the public (cf. Franklin 1999b). The first licences for freezing embryos were issued in August 1991 when the HFEA was established; the first ran out in August 1996.

Dolly was born in 1996, but the issue of her birth only publicised in 1997. She was cloned from a ewe's udder cell, which had been fused into a de-nucleated egg cell and then implanted into a surrogate mother sheep. The sheep was created by Ian Wilmut and his team at Roslin Institute, Edinburgh. The goal of the technique used to create Dolly, which is referred to as 'cell nuclear replacement' (CNR), was the engineering of transgenic animals that carry human genes.[26] While cloning in livestock breeding was not a new practice overall, what was different in Dolly's case was the fact that she had been cloned from an adult cell—a technique thought to be impossible. Created by the same team of scientists, Dolly's predecessors Megan and Morag, born in July 1995 (The Observer Review, Mar 10, 1996), were clones as well. However, they were cloned from embryo cell nuclei, which were fused into unfertilised de-nucleated egg cells. Megan and Morag had already triggered a debate over pros and cons of animal cloning and the possibility of human cloning in the British press. Carried out in order to find ways of easier genetic manipulation of animals in which genes could be altered before implantation, these experiments aroused fears of human cloning and human genetic manipulation being possibly closer to realisation. UK newspapers were full of horror stories of cloned dictators, such as Saddam or Hitler; of cloning as a source for 'spare

25 For a discussion of the shift in focus from 'animal cloning' to 'human cloning' in British media debates cf. Franklin (1999a).

26 Indeed, after Dolly the Roslin team created a transgenic sheep Polly, which carried a human gene for the production of Factor IX in her milk. Factor IX is used as a treatment for haemophilia.

parts' for transplantation; the possibility of reproduction without male involvement,[27] and the likelihood of a future technique that allows mammalian cloning from an adult cell. However, most newspaper reports were about farming and livestock breeding, reflecting the purpose of the creation of Megan and Morag. While scientists and commentators admitted that it was possible to clone human beings, most said that it was neither practical nor ethically acceptable. Ian Wilmut, one of the creators of Megan, Morag, and Dolly, said in 1996: "I cannot see why anybody would do such a thing. […] That would be an appalling prospect" (as quoted in The Times, Mar 8, 1996a).

In February 1997, the topic reappeared on the front pages—this time, its name was Dolly. Several actors, including the then president of the Royal Society Sir Aaron Kluge, highlighted a potential loophole in legislation that may allow the cloning of humans. As it did not involve an embryo, which, according to the law, is the result of fertilization, the method used to create Dolly was not explicitly ruled out by the HFE Act, and thus human cloning could become reality. However, the HFEA insisted that no embryo can be created without a licence; as quoted in Daily Telegraph (Feb 26, 1997) a HFEA spokesman asserted: "We would take control over it." Although many commentators still opposed human cloning as ethically unacceptable, others were concerned about the "atmosphere of criticism" (Ian Wilmut as quoted in The Guardian, Feb 26, 1997) surrounding the experiments at Roslin Institute. A few days after Dolly's birth was announced, the Ministry of Food, Fisheries, and Agriculture (MAFF) declared that the funding for the Roslin Institute would be halved for the next six months and withdrawn in April in 1998. MAFF said that the reduction and termination of funding was justified because "the technique for cloning sheep is established and funding has achieved its aims" (STC 1997b, 48; cf. Daily Telegraph, Mar 1, 1997).

In the first few months after the announcement of Dolly's birth it was the centre of media attention. All national newspapers covered the story, filling front as well as science and opinion pages. In the beginning, the

27 The absence of male sperm in cloning seemed to be the most striking problem of cloning for many reporters. One went as far as to claim that "American feminists, who are the vanguard army of the nice in the war against the good, have long resented the part played by men in reproduction. Having successfully deprived the foetus of its rights, they can scarcely stop short of their goal, which is to make men redundant" (The Times, Mar 8, 1996b).

general theme in the media was *human cloning*, which—corresponding to the Dolly method—referred to the production of a human being via cell nuclear replacement.[28] Although media attention lessened after a while, until the end of 2000 the topic of human cloning recurred intermittently in the context of different, more or less related issues and events.[29] Especially engaging or establishing official bodies that deal with the issue and also other events, such as Richard Seed's announcement to build a cloning clinic in the USA in January 1998, were commented on and linked to a discussion on cloning.

In March 1997, when the Science and Technology Committee (STC)[30] held a consultation session with witnesses from the HFEA and the Roslin Institute, a discussion of the need to balance potential benefits of cloning against its shortcomings appeared in the press. The Times quoted Ruth Deech's (then chairwomen HFEA) statement at the STC consultation: "Although the prospect of producing fatherless identical human beings was repugnant to many people, the potential benefits should not be ignored and they would never be realised if a total ban on such research were imposed" (The Times, Mar 6, 1997). Many press commentaries maintained that it was essential to retain flexibility in the law in order to accommodate future scientific developments and to prevent cloning experimentation on humans.[31]

Especially in the first few weeks after the announcement of Dolly's birth, many different topics and aspects, which can be summarized under the umbrella issue of *the relation between science and society,* hit the papers and were debated. Newspaper articles discussed the regulation of scientific inquiry, scientists' responsibility, science 'ethics', as well as the need of public debate on the issue. Nobel Peace Prize Winner Joseph Rotblat stated that the prospect of human cloning was "science out of control" (as

28 The term 'human cloning' refers mainly to what at present would be called human *reproductive* cloning. At the time, commentators did usually not differentiate between what is currently called 'reproductive' and 'therapeutic' cloning. While some of the so-called benefits of human cloning were later depicted as 'therapeutic cloning', the terms and distinguishing between 'reproductive' and 'therapeutic' cloning came into existence, when they were introduced in the consultation by the Cloning Working Group of the HFEA and the HGAC in 1998, which will be discussed in chapter 3.6.

29 For an analysis of newspaper coverage cf. also Kitzinger et al. (2003) and Wilkie (2001).

30 This consultation will be discussed in more detail in the next chapter.

31 Cf. for example: The Independent (Feb 28, 1997); Daily Telegraph (Mar 6, 1997); The Times (Mar 6, 1997); Financial Times (Mar 6, 1997).

quoted in The Guardian, Feb 26, 1997) and that "scientists must take responsibility for their work" (as quoted in Evening Standard, Feb 28, 1997). Others said it was not right to stop scientific curiosity and scientific research (The Independent, Feb 28, 1997); that the technology was "out there" and somebody would make the breakthrough (Wilmut as quoted in The Guardian, Feb 26, 1997), adding that "ethics will have to cope" as science advances (Daily Telegraph, Mar 5, 1997).

Many commentators demanded a public debate on the issue, a number of them complaining that decision-making on science issues was still restricted to experts.

> "Hitherto, harassed politicians have been content to shift the decision-making to panels of 'experts' and rely on high minded academics to tell us what we ought, or ought not, to do. It is time to bring a bit more democracy into the process, to encourage the widest possible public debate, and to find out what we all think the rules should be" (Daily Mail, Feb 24, 1997).

At Westminster, MPs called for the examination of cloning and genetic engineering by a special House of Commons committee. A motion tabled by conservative MP Bill Cash and backed by about 35 MPs from all parties called for an *ad hoc* select committee to discuss the ethical issue arising from the developments. Cash said: "We cannot have these matters decided by scientific advisory committees" (as quoted in Daily Telegraph, Feb 28, 1997). Liberal Democrat David Alton called on the administration to halt cloning experiments and start a public debate on the issue (The Times, Feb 27, 1997). Alton urged the appointment of representatives to the Human Genetics Advisory Commission, "who reflect traditional concerns about the sanctity of human life, [...] and who disagree with the Nuffield Council's[32] view that the species boundaries, in any case, are not inviolable" (as quoted in The Times, Feb 27, 1997). However, it was not necessarily those who disapproved of the developments that called for public debate but also supporters. Ian Wilmut made an appeal to the public in defence of his work: "It is now up to society to decide how […] [this technology, slh] should be used and we welcome any discussion of these matters" (as quoted in The Guardian, Feb 26, 1997).

32 The Nuffield Council of Bioethics is an independent ethics board that was established in 1991 as an initiative of the Nuffield Foundation. It receives funding from the Medical Research Council.

Actors engaged in the debate also did not agree on the legal aspects of the matter. While several spoke of a 'legal loophole' within the HFE Act, others maintained that human cloning was definitely forbidden in the UK. The British government had stated that UK law did not allow human cloning and the HFEA had argued that the British system of control was tight enough to prevent it. The ambiguity arose from the fact that it was unclear whether or not the result of the fusion of egg-shell and cell-nucleus was an embryo according to UK law, thereby falling under the auspices of the HFEA. The HFE Act defines an embryo as the result of *fertilization*, which did obviously not take place in the production of Dolly. Dolly was engineered from an adult cell and an egg cell and hence without the involvement of an embryo as defined by the law. The existing ambiguity, however, was strong enough to make official inquiry necessary.

Three official bodies were engaged with the issue and provided statements and recommendations for the government: the House of Commons Science and Technology Committee (STC), the Human Fertilisation and Embryology Authority (HFEA) together with the Human Genetics Advisory Commission (HGAC) conducting a public consultation with about 200 responses, and a newly established Expert Advisory Group on Therapeutic Cloning directed by the Chief Medical Officer (CMO) Liam Donaldson. Official actors were predominantly concerned about public anxieties towards scientific experimentation in general and towards human cloning in particular. What followed were public as well as expert consultations and hearings that centred on questions of regulation, ethics, public concerns and public trust, risks and benefits of cloning research, etc. The following chapters analyse these procedures and statements in terms of how they defined the problem, what kinds of solutions they suggested and how these changed throughout the years.

3.5 Public Concerns versus Scientific Development: Loss of Public Trust

In the UK, high-ranking policy-makers were involved in the debate from the very beginning. The first official body that addressed Dolly's birth and the possibility of cloning humans was the House of Commons' Science and Technology Committee (STC). The STC is a continuous advisory body that inquires into science and technology issues, including human genetics, under the auspices of the Office of Science and Technology (OST). It consists of eleven MPs, and the party balance mirrors that of the House of Commons. Appointed in July 1992, its task is to examine the OST's as well as the associated bodies' expenditure, administration, and policy. The STC determines its own programme and regularly publishes reports on findings and recommendations.

After the Roslin Institute announced Dolly's birth, the STC immediately decided to consult on the issue. On 5th and 6th March 1997, the committee interviewed members of the HFEA and of Roslin/PPL Therapeutics Ltd.[33] Also included in the consultation were memoranda and letters by the Home Office; the Department of Health (DoH); the Institution of Professionals, Manager, and Specialists/Biotechnology and Biological Sciences Research Council (BBSRC) Branch—the Roslin Institute Section; the British Medical Association (BMA); the Office of Science and Technology (OST); the Ministry of Agriculture, Fisheries, and Food (MAFF); the Constitutional and Community Policy Directorate within the Home Office; the Institution of Professionals, Managers, and Specialists (IPMS); and the Roslin Institute together with PPL Therapeutics. A memorandum by the Church of Scotland and a letter and attachments from the Genetics Interest Group (GIG)[34] were also presented to the House, however, they were not included in the final report (STC 1997a, xxi-xxiii). The statements and proceedings of the consultation will be the focus of the next chapter, which will investigate how the different actors, witnesses in the consultation, and the STC itself interpreted the problem and how they proposed to solve it.

Problematisation: Why a Consultation?

In its Fifth Report of Session 1996-1997 *The Cloning of Animals from Adult Cells*, the STC (1997a) was most concerned with the scientific challenges of the work at the Roslin Institute and the benefits it could bring. It also fo-

33 PPL Therapeutics Ltd (formerly Pharmaceutical Products Ltd) is one of the world's leading companies in applying transgenic technology for the production of human proteins. It was founded in 1987 to commercialise the transgenic technologies developed at the Roslin Institute. PPL Therapeutics holds a licence to exploit the work in this field. Cf. http://www.ppl-therapeutics.co.uk/welcome/welcome.html, Jan 20, 2006.

34 The Genetic Interest Group is an umbrella organisation of about 120 charities that support people affected by 'genetic disorders'; cf. GiG webpage at: http://www.gig.org.uk/members.htm, Oct 29, 2003.

cused on questions regarding the adequacy of the law applicable to cloning and on related issues involving both animals and humans. Additionally, the session deliberated on the future of the Roslin Institute (ibid., v). However, what the STC found most problematic was that the 'strong public reaction', especially in newspapers, had overshadowed the discussion of benefits. The report recommended that, although public concerns were derived from 'irrational fears', they still had to be recognised and responded to. Otherwise, "there is the danger that public opinion will block benign developments that on mature consideration are found to be humane and acceptable to all" (ibid., ix). Likewise, the OST stated in its memorandum to the consultation that "hand in hand with scientific achievements and its potential applications come equally significant issues of appropriate regulation and public confidence" (OST 1997, 43). Public reactions to the announcement of Dolly's birth and their possible negative effects on scientific development were interpreted as highly problematic. Ways had to be found to deal with public concerns and to prevent scientific development from being blocked. Indeed, a problematisation was presented that can be described as a *conflict between public concerns and scientific development.*

To a considerable extent, both the examination of witnesses and the memoranda addressed the issues of *public concerns*, *public interest*, and *public debate*. The problems of how to tackle or prevent public anxieties, what was designated as ill-informed public debate, or understanding the relation between scientific experimentation and public debate were discussed at length, as well as also the necessity of advisory committees in order to ease or facilitate this relation. Prominent issues in the consultation were, among others,[35] the possibility and (questioned) desirability of applying the Roslin technique to human beings; the question of public funding, especially the reduction and termination of MAFF funding of the Roslin Institute; the pros and cons of tight regulation; prohibition of cloning humans; as well as the role of 'ethics' in the decision-making process.

Public concerns and public debate and the (potentially restrictive) impact on scientific developments thereof were regarded as the most problematic aspects necessitating response. However, the list of consulted groups and individuals hints at a problematisation that interprets the problem as one that can be dealt with within the existing system in terms of specialist knowledge and expert advice: Problematisation of public con-

---

35 For example, the regulation of animal experimentation and agricultural aspects of biotechnological experiments at Roslin, which will not be discussed in this study.

cerns and public debate framed the issue as a matter for specialists of science and technology, of finances and economics, and of the regulatory and advisory system.

Public Concerns and Scientific Development

While the discussion of 'public concerns' formed an important part of the examination, these concerns were only vaguely portrayed.[36] They were described as 'unreasonable', 'nonsensical', and 'immature' and as deriving from false information, from ignorance, and general disinterest of the public and the media in scientific experimentation. Nonetheless, public concerns were taken seriously and ways were sought to reassure public trust. However, the concerns themselves were neither deliberated on nor negotiated. It was decided instead to counteract them through education, persuasion, and regulation.

The public debate was considered problematic in itself, and as a potential hazard to scientific development. It was considered to be unorganised, misinformed, and driven by irrationality and, therefore, a problem. Witnesses in the consultation stated that the anxieties within the public resulted from "unreasonable" and largely "nonsensical" conceptions, which were published in the press or brought forward by "eminent scientists", associating the Roslin animal experiment with humans (Wilmut in STC 1997b, 26). Previously, Ian Wilmut had already expressed his irritation in a newspaper interview about the "atmosphere of criticism" surrounding his work (The Guardian, Feb 26, 1997). This led the OST (1997) to suggest that educating the public was (part of) the solution. The OST found it particularly important to present scientific information in a clear and accessible form so as to foster an informed public understanding of science. Therefore, the OST stated in its memorandum that they would be able to "provide an understanding of scientific facts and processes as well as an appreciation of what benefits are offered and what cannot yet be achieved" (ibid., 44). The goal, however, was to enhance "the UK's lead in Europe in the industrial exploitation of biotechnology and for the UK to outpace

36 Williams et al. (2003) describe a similar phenomenon for media reporting on 'ethical' debates on the embryo: Whereas 'concerns' were recognized as existing, they were not reported themselves. Upon studying the newspaper coverage, one gets the impression that articles and comments respond to some concerns or critiques, which were, however, not mentioned themselves.

international competitors" (ibid., 44), which was understood to be at stake because of public mistrust. Focusing on competitiveness, the OST continued: "the Crusade [for Biotechnology; slh] nevertheless recognises the importance of public confidence" (ibid., 44). Thus, already on the outset, for the OST it was quite clear that the 'UK's lead in biotechnology' was not to be challenged, but, indeed, the aim of the whole endeavour was to take a lead in biotechnology. The 'problem' was the lack of public trust that might endanger achieving this aim.

As public reactions to and debate on scientific development were problematised, it is not surprising that questions of how to conduct and organize public debate in relation to scientific development came on the agenda. The option of a public debate *before* scientific experimentation takes place was not embraced by Wilmut and his colleagues, who stressed the unpredictability of the results, which, in their view, would make it impossible to "carry out an effective assessment before a project is begun" (Wilmut in STC 1997b, 24). Additionally, scientific publication rules prevent any details being disclosed before journal articles are published and thereby hinder the early release of information into the public domain: "We got this research into the press and into the public debate as soon as we possibly can", said Graham Bulfield, Director and Chief Executive of the Roslin Institute (STC 1997b, 20).[37] Nonetheless, even if they desired it, early public debate on science would be difficult to launch, because "until something like a lamb comes along then really quite frankly the public and the media is not interested" (Bulfield in STC 1997b, 25). However, particular applications of human cloning, as Bulfield (ibid., 26) stated, "are issues for society, they are not issues for scientists. [...] I do not think that our opinion on it is any more valuable than anybody else's". In another instance, he said that "it is not my view that counts, it is the view of the public" (ibid., 25). In spite of this, whether or not the public should be consulted or have the power of veto over certain experiments before they are carried out was considered problematic. It was stated that the public was unable to name "uses and abuses" until experimentation was successful (Colman in STC 1997b, 26).

To sum up: On the whole, scientific experimentation was framed as generally acceptable and questions of unacceptability were conceptualised according to 'uses and abuses' of scientific results. These, however, can

37 On the publication politics of the Roslin Institute concerning Dolly cf. Wilkie (2001).

only be assessed *after* experimentation has been successful. The development of science itself was not on the agenda. Public responses, on the other hand, were problematised in terms of the inability of the public to oversee scientific research and development. The witnesses of Roslin/PPL called for a committee to address or prevent public concern and mediate between scientists and the public "so there could be an informed discussion in a debating format [...] whereby many of the public interests were represented" (Bulfield in STC 1997b, 25). Bulfield said that the public might have been reassured if an advisory committee had permitted the research and if the public had been informed that the Roslin Institute had obtained the required permission to perform the experiments. The framing of concerns as the result of ill-informedness that generated a deficit of trust led to the presentation of a solution that entailed an organised, informed, and open expert-advised public debate that would ease public concerns.

Regulation and Public Trust

A key aspect suspected of causing public mistrust was a *regulatory uncertainty* of some kind or other within the public. Thus, the solution considered most effective in addressing the problem of 'public concerns' was a 'tight' regulatory system: "As we have said before, the best way to ensure public confidence is by having a regulatory system in place which can deal with both the ethical and the practical issues which will inevitably arise" (STC 1997a, x).

In the examination of HFEA members, the reassurance of public trust by prohibiting cloning outright was weighed against possible future benefits of the technology. Prohibiting human cloning on the whole, a measure that would "indeed reassure the public" (Deech in STC 1997b, 4), was dismissed because it impeded possible future benefits. Instead, the HFEA suggested that concerns be met "by clarifying and, if necessary, broadening the definition of 'embryo'" (ibid., 5). However, it was left to parliament to decide for one or the other. Broadening the definition of 'embryo' to include embryos that were produced with the Dolly technique within the realm of the HFE Act was understood to potentially address and reduce public concerns. Public concerns were framed as reactions to regulatory uncertainties rather than to other factors and were addressed as such. On the other hand, witnesses set store by the assurance of being able to conduct scientific experimentation in a relatively unconstrained manner. Thus, although regulation was deemed important, it had to be flexible enough to not impede research. As the OST claimed:

"It is essential that our advisory and regulatory system keeps pace with developments in research. The system needs to be flexible enough to respond to new developments without presenting undue burdens to innovation and potential advances in human health and agriculture. *It needs to reassure the public without damaging the quality of the scientific research output*" (OST 1997, 43, emphasis added).

Nonetheless, the examined members of Roslin/PPL made sure that, despite human cloning being possible, "it is the unanimous view of the group within the Institute and the Company that we would find this sort of work with human embryos offensive. [...] We would support wholeheartedly the idea of prohibition in as effective a way as possible" (Wilmut as quoted in STC 1997b, 23). Notwithstanding this clear statement, Bulfield said that "in the future something might happen that would demonstrate that this technology might be useful but we cannot see that at the moment" (as quoted ibid., 23). Therefore, Wilmut suggested to leave the matter to the judgement of the HFEA instead of banning experimentation altogether. "If you start covering up little holes in the legislation as the technology moves on we can not predict if something else comes along that might again find a hole in it" (as quoted ibid., 24). Legislation, it seemed, was not considered to have the task of 'covering holes' and thus preventing unacceptable scientific research but to facilitate its unconstrained expansion—or, as Irwin had said, to secure science's "licence to practice" (Irwin 2006, 308). Likewise, HFEA member Ruth Deech clearly wanted to foreclose an outright ban on human cloning. Although "we would never want to license any treatment that might be risky" (as quoted in STC 1997b, 4), what appears as risky today might be beneficial tomorrow: "after all, science goes on and things that seemed awful a few years ago may prove to have beneficial effects in a few years' time and we would not want to stop that" (ibid., 4). To avoid 'damaging' this kind of research, Deech suggested that the definition of an embryo be broadened in the HFE Act in order to cover human embryos produced via the Dolly technique, thus making them subject to the licensing of the HFEA. An absolute ban on human cloning would mean that "you will probably have to come back to the Act every few years and change it" (ibid., 8).

Apparently, examined actors put much effort into preventing a comprehensive ban on human cloning. The problem they described was that public mistrust and 'uninformed' public debate could block research and possible future benefits. The pursued goal was to introduce and maintain flexible legislation that would easily adapt to new developments and, at the same time, reduce anxieties within the public and reassure public trust. The STC report recommended that the HFE Act be amended to assure the prohibition of human cloning while securing and maximising the UK's research potential (STC 1997a, ix). A positive step in achieving this was broadening the definition of 'embryo'. In this case, the HFEA would gain authority over licensing this particular research and could thus review and decide on each research application individually. Human cloning would thereby fall under a comprehensive licensing system, and legislation would be flexible enough to allow research projects that the HFEA considers worthwhile. It was expected that this strategy would, at the same time, pacify public concerns.

The Case for Science

The reduction and, finally, termination of MAFF funding of the Roslin Institute within the next year was another key topic in the examination of Roslin/PPL witnesses. The MAFF had written in its memorandum that "it is time for industry to take over" (MAFF 1997, 48). Roslin witnesses claimed that the reduction and termination of money flow at such short notice would jeopardize particularly the work of Ian Wilmut's team (Bulfield in STC 1997b, 26). Unstable public funding, so it was claimed, put support from other, commercial funders at risk. While witnesses admitted that the MAFF legally had the right to discontinue funding, they argued that it would be against the interests of the UK and the wrong decision to make. It was suggested that the ministry had the duty to support "long-term strategic science", and this "applies over a wide range of industries" (ibid., 27). An interim between scientific accomplishments and their predictable success for industrial and commercial exploitation was often required. In the case of Dolly, the technology was not "near enough the market yet that the animal breeding industry is capable of taking it up" (ibid., 28). Witnesses also made clear that jobs were at stake: Bulfield said: "I am going to have to make a significant number of our permanent staff redundant" in order to "protect this work" (ibid., 28).

Altogether, MAFF's decision was problematised as jeopardising the British science base as a whole, as scientists might 'have to leave the country' if the flow of public money is disrupted.

"Now we have a lead for this technology in this country at the moment. I am not doing this for the benefit of my colleagues or for the benefit of the Roslin Institute; we are supposed to be doing this *for the benefit of the UK* and I want to see this technology exploited in the UK, not elsewhere, but my major aim in the next few months is to keep Ian's [Wilmut; slh] team together. I do not want somebody coming along with a blank cheque and poaching this guy and taking him to California. That's not what I want to happen" (ibid., 27, emphasis added).

The scientific and economic future of the UK was said to be at stake if public funding of Roslin and other BBSRC[38] institutes was stopped. The BBSRC stated in its memorandum:

"Questions about stability are central to the future health of the UK biotechnology and agricultural sector and the industries that they underpin. The next industrial revolution will be dominated by biology. [...] Thus, it is essential for the nation's future economic well-being that the UK has a well-funded research base, staffed with well-motivated scientists with appropriate support and facilities" (BBSRC 1997, 40f).

Thus, MAFF's decision to stop the funding of research into cloning sheep was represented as not only putting Wilmut's team and Roslin staff at risk but also the British science base and the nation's economic future. The reductions would, the STC report stated, damage the Roslin Institute and "the Science Base as a whole, and British industry" (STC 1997a, xvi). The problem was not simply a problem of the Roslin Institute but a problem for the whole British nation, its economy and industry. Hence, it was not surprising that the STC recommended that the reduction of MAFF funding should be reconsidered or otherwise MAFF should bear the costs of its actions. The STC should be given the authority to conduct an inquiry into MAFF's commission and advisory procedures. The discussion of MAFF's decision was used to make the case for British science. Not only the science sector itself but, indeed, Britain's economic future was said to be at

---

38 The BBSRC, that is, the Biotechnology, and Biological Sciences Research Council, is the biotechnology branch of the Institution of Professionals, Managers, and Specialists (IPMS). The BBSRC is a public funding agency that supports academic research and training in non-medical life-sciences at universities and institutes throughout the UK, including research at the Roslin Institute (cf. BBSRC wepapge at: http://www.bbsrc.ac.uk; Sep 2, 2003)

stake if the funding of cloning science was stopped. Scientific development was portrayed as being in national and public interest and, in fact, not problematised.

Benefits versus Risks

Another important frame was that of *benefits and risks.* Cloning, the STC stated, "faces us with moral dilemmas", which occur "when risks and benefits have to be balanced" (STC 1997a, vi). However, discussion of science was "overly concerned by potential risks, and has not properly considered the potential benefits" (ibid.). Interestingly, portrayal of the so-called and often mentioned benefits of the research remained very vague throughout the examination of witnesses as well as in the final STC report. Often only the term 'benefits' occurred without much elaboration. If 'benefits' were elucidated, their representation was rather broad, often imprecise, and always conditional. Actors spoke of the *potential* or the *great promise* of the research—of uncertainties rather than certainties. For example, the STC report referred to "possible" medical treatments, such as "the *implications* that [the research; slh] […] *can have* in the way of regenerating a failing heart" (ibid., vii, emphasis added). The report concludes that "We too believe that research *should* relieve suffering and improve the health of individuals and human kind as a whole, and have no desire to restrict it unduly" (ibid., viii, emphasis added). Witnesses spoke of "beneficial effects" (Deech in STC 1997b, 4) or of "realistic targets and objectives", which would include "congenital and other examples of human diseases" (Roslin Institute/PPL Therapeutics 1997, 12). The Department of Health emphasised that the "positive therapeutic role for research in this area" (DoH 1997, 32) could be affected by a total ban of cloning technology. A ban, the DoH claimed, could "prevent research into a number of crippling or life-threatening inherited illnesses in humans [...]" (ibid.). Problematisation in terms of a *moral dilemma between benefits and risks* helped to portray the issue as a question of either pass or fail: stop or restrict scientific development and thereby lose the possible benefits for humankind. At the same time, the extensive use of conditional phrases betrayed great uncertainty about the goals and future results of scientific endeavour.

Ethics and Public Trust

Also the use and function of the notion of 'ethics' seem noteworthy: In the consultation exercise 'ethics' had, on the one hand, the function to *signal that public concerns were taken into account.* On the other, 'ethics' meant the *balancing of risks and benefits.* As described above, addressing 'public concerns' or 'interests' was understood as necessary because of the public's (alleged) power to block research rather than because it was an issue of democratic deliberation. Public concerns had to be counteracted, so it was assumed, through comprehensive but flexible regulation and structured information, which reassure public trust in science without endangering its development.

However, references to ethics remained rather technical and arbitrary. Queried on the role of ethics within the regulatory and licensing procedures of the HFEA as well as on whether ethics inhibits licensing in any way, Ruth Deech explained that the HFEA has the power to, and on occasion does, consult the public and regularly consults the science profession. It also engages in advising patients, encouraging them to deal with ethical concerns (Deech in STC 1997b, 4). Additionally, the HFEA had set up an ethics committee and makes sure that the sexes as well as different professions—including medical, theological, and philosophical professionals—are balanced, Deech stated. She concluded that "this infects our work" (ibid., 4).

The memorandum submitted by the British Medical Association (BMA) particularly focused, so it was claimed, on the "ethical implications of the possible use of such technology in humans" (BMA 1997, 41). The BMA asserted that it was especially concerned with the possible harm that could arise from cloning humans. As harmful outcomes that could possibly occur, the BMA named the birth of identical children at different times and the unforeseeable psychological consequences on the children, the possibility to store clones in order to provide identical siblings for tissue transplantation, or the production of a 'perfect' replacement of a dead child. While such occurrences could be prevented by law, the BMA claimed that the fundamental moral concerns remain. Cloning humans could "undermine the intrinsic value of each person, dilute individuality and reduce the genetic variation on which species depend for their evolution and survival" (ibid., 41). Therefore, the BMA "currently" opposes any cloning techniques to be applied to humans (ibid.). However, "as with all of the BMA's ethical guidance, this position will be kept under review but it is unlikely to change in the foreseeable future" (ibid.).

On the whole, 'ethics' was portrayed as a device that dealt with and managed the problems and conflicts thought to be surrounding biotechnical research in general and the Dolly technique in particular. It was presented as a means of balancing risks and benefits and preventing the abuse of science, on the one hand. On the other, it was introduced to signal that public 'irrational' fears and 'immature misunderstandings' were taken into account, that these were being responded to. Consequently 'ethics' was a means to manage public concerns in addition to, finally, gain public trust in science. This was made possible by interpreting the 'problem' either as a moral dilemma between risks and benefits or as a problem of public concerns, which, especially in the press, overemphasised the risks and displayed ignorance towards the benefits. In this context, 'ethics' was presented as a means to restore the 'right' balance between risks and benefits and, on these grounds, re-establish public trust.

In their final report, the STC refers to *Nature* magazine, which had called for a moratorium on cloning a week after the birth of Dolly had been announced. A moratorium seemed the perfect solution to respond to (ethical) concerns without putting scientific progress at risk. As stated in *Nature*,[39]

> "The experience with genetic engineering technology shows that an explicit moratorium can be productive. On the one hand it sends a public signal that ethical considerations are important and need to be taken into account, and allows that to happen. On the other, the fact that a moratorium is, by definition, likely to be lifted at some point indicates that further development is anticipated" (as quoted in STC 1997a, x).

Just like regulation, ethics—symbolised in a moratorium—was considered to be a tool to reassure public trust so that science could advance without obstruction.

Reassure Public Trust But Do Not Stop Research

The issues discussed in the STC consultation were manifold. Nonetheless, the major concern was that public mistrust in and anxieties about science could block the development of science. Accordingly, recommendations in

39 Nature, Mar 6, 1997, Vol. 386, 1.

the report concentrated on problematisation of *strong public reactions, which were characterised by the concerns overweighing the benefits*. The recommendations made by the STC reflected several of the assumptions on the nature and origin of these concerns: In line with the assumption that public concerns derived from uncertainty about regulation, the STC report recommended the amendment of the HFE Act through primary legislation[40] by broadening and clarifying the legal definition of 'embryo'. The re-definition of the term 'embryo' sought to include all cloning techniques, to allow the HFEA to decide on each case individually, to retain the subtleties of the act, and to allow potential benefits to be pursued. "The HFEA could then be given the discretion as to whether work using the technique at Roslin, or other comparable techniques which might be developed in the future should be banned" (STC 1997a, xiii). Human reproductive cloning, however, should be banned. The committee also expected the Human Genetics Advisory Commission (HGAC) to play a significant role in advising parliament on the issue. Nevertheless, the STC stated, "evidence we have received suggests that the regulatory system in place is already sufficiently strong to ensure that there is no need to panic" (ibid., ix).

With regard to the problem of ill-informed public debate, the STC said that it "wished to examine issues which appeared of current importance and to place evidence on the record which should be of use for the [Human Genetics Advisory; slh] Commission in its further deliberations" (ibid., v). As a response to public concerns that might endanger scientific 'progress', the STC suggested more expert advice and expert-organised public debate. In line with comprehending the problem as concerning public anxieties that arise from 'irrational' fears, which are based on misunderstandings and ignorance of the so-called benefits as well as regulatory and 'ethical' uncertainties, the two expert bodies—HGAC and HFEA—were considered to be the adequate bodies to deal with the issue. The STC report recommended that the HGAC, that is, the official body for advising on social, ethical, and economical consequences of the new genetics, should play a significant role in analysing the issue further and give advice on whether to remain, extend, or replace the ban on cloning. The HFEA, that is, the official regulatory/licensing instance, should be left to decide on new projects (ibid., xiii). The decision-making process was therefore to be left in the hands of expert bodies.

---

40 However, when it came to the parliamentary decision, the government introduced secondary legislation. Cf. 3.8.

The Government Response

In December 1997, the British government responded to the STC report (DTI/OST 1997). Even though it agreed to most of the statements and recommendations, it also saw the need to consider further whether or not and how legislation should be amended. The government affirmed that the UK play a leading role in science and technology, which, as the response stated, had been proven and furthered by successfully cloning Dolly. According to the government, however, this research bears ethical implications—including "significant new benefits in research in human and animal health" (ibid., 1). The response added that the "deliberate cloning of human individuals" is regarded as "ethically unacceptable", whereas the use of cloning techniques, "where research is being carried out on serious inherited illnesses and where the end result will not involve the cloning of human individuals", is considered acceptable (ibid.). However, it emphasised that "the issue of human cloning is not straightforward" and that the government accepts "that research into some serious diseases can take place using cloning techniques" (ibid., 3).

On this issue, the administration outlined that its own responsibility was to "foster excellence in United Kingdom genetics research" and to "ensure that public confidence is maintained" (ibid., 2). It acknowledged the "acute public sensitivities", thereby pointing out that the HGAC and the HFEA offered consultation, advising on ways to "build public confidence in, and understanding of, the new genetics" (ibid., 4).

> "The Government fully recognises that the strength of the United Kingdom's science base made possible the breakthrough at Roslin. The cloning of Dolly from adult sheep cells aroused substantial public interest and concern. *The Government must maintain public confidence in the future handling of such cases. The work of the HFEA and HGAC will go a long way to ensure that public debate is focussed on the appropriate areas.* Meanwhile the Government will continue to foster excellence in UK genetics research as it has exciting potential to bring about benefits in areas such as health and agriculture, and thus improve the quality of our lives. The Government will also strive to *ensure that the debate on the ethical issues keeps pace with advances in these technologies*" (ibid., 6f, emphasis added).

In its concluding remarks, the government gets to the point: While it considered genetic research to be vital for the UK's competitive strength and worthy of being promoted in order to improve the 'quality of life', public concern posed obstacles to its uninhibited progress. Well aware of the political risks that public mistrust in political decisions might imply, the administration claimed that public confidence must first be reconstituted. The 'problem' of cloning was not a problem of science but a problem of public mistrust. For that reason, the administration wanted to make sure that "the debate on ethical issues" was reorganised in order to "keep pace with advances in these technologies". It agreed with the STC and presented the HFEA and the HGAC as being the adequate bodies to address the issue: It is the task of the HGAC to address and advise on social, ethical, and economical issues arising from human genetics. The HFEA "acts as a safeguard for public concerns" (ibid., 2).

Summary: Science's Licence to Practice

The STC consultation took place two weeks after Dolly's birth had been announced, at a time when debates in the press had only started. However, official bodies as well as scientists were immediately concerned about 'public anxieties' towards scientific developments such as 'human cloning'. Thus, what was defined as problematic in the first phase was the relationship between public concerns, scientific development, and UK economy. Different interest and stakes were introduced, including funding and survival of Roslin and its staff, the UK economy, and the general 'necessity' to continue scientific experimentation. UK economy and competitiveness in the biotechnology sector were presented as legitimate and unquestionable 'goods' that depend on science and technology advancement. The nation's well-being (or 'health' as far as the government was concerned) was declared to be at risk if scientific development—especially cloning research—was limited. On the other hand, concern was voiced that public apprehensions or mistrust would pose obstacles to and may even block scientific development and endanger the UK economy and the well-being of the nation as a whole. These concerns were presented as being the issue of ill-informedness and immature public debates, which overemphasised risks rather than benefits of scientific research. Thus, actors promoted a new form of 'more mature' debate and stressed the need for educating the public to enable it to participate in an 'informed' and more rational debate. However, the report did not mention or quote any of these 'public concerns' that were presented as problematic. The answer was, nonetheless, to promote a more rational debate led by experts and to follow expert advice. Because the representatives of the 'concerns' were absent, what the report was talking about or who it was arguing against remained obscure. Therefore, interpretation of 'concerns' remained exclusively in the hands of the group of experts present in the consultation as well as in those of the STC itself. 'Omitting' opposing statements, however, was less an *omission* than an *exaggeration* and diversion: Even though there was some opposition to or critique of 'human cloning' in these first two weeks (before the STC consultation took place), on the whole there was much more emphasis in newspaper coverage of the new research on the *benefits* and much less discussion of the risks involved or concerns raised (cf. also Kitzinger/Williams 2005; Wilkie/Graham 2001). It seems that the STC addressed a potential opposition rather than a clearly identified one.

Still, problematisation of concerns as deriving from 'misinformation' or 'mistrust' was a powerful tool for disqualifying potential opposing arguments and can be interpreted as an influential framing device, which not only channelled the discourse in a certain direction but also emphasised expert authority in this field while excluding other 'misinformed' participants. To use the words of Alan Irwin (2006), next to a (traditional) deficit model of the public, that is, next to interpreting public scepticism towards science as the result of misinformation and misunderstanding, what we see here is the application of a new deficit model: It is that of a deficit of public trust, which formed the basis of the overall problematisation of 'public concerns that might block scientific progress' and which needs to be tackled in order to create a 'progressive' environment for science. As a response, apart from suggesting that 'the notion of an embryo in the law be redefined' and thereby 'reassure trust in regulation', a public debate was recommended. It was to be guided by experts. The 'ethical debate', however, as the government wrote, should 'keep pace with advances in science' and not vice versa: it is science that should take (keep) the lead in the debate, leaving no room for a serious political intervention in science. To use the words of Alan Irwin again, we can indeed ask: "What scope can there be for dialogue when the direction is already set", (ibid., 316) that is, when scientific 'progress' is the unquestionable goal? At a time when debate had not yet begun, exaggerating oppositional statements and problematisation of public concerns were means of underpinning calls for a 'new' public debate—a debate that should be characterised by knowledgeability, rationality, and expert guidance. They were a means of stimulating public debate that 'keeps pace with science' and does not challenge it.

3.6 Let's Consult the Public!

It is not surprising that the next stage in the debate was to establish a working group of the Human Fertilisation and Embryology Authority (HFEA) and the Human Genetics Advisory Commission (HGAC), the two bodies that had been portrayed as the adequate bodies to address the issue of 'public concerns'. The HGAC had been established in 1996 in order to counsel the government on ways to build public confidence in and understanding of the new genetics as well as to review developments in this field and advise on social, ethical, and economical consequences arising from them.[41] It is an independent advisory body charged with setting its own priorities, but it can be requested to provide urgent advice to ministers. In 1999 it was replaced by the Human Genetics Commission (HGC).[42] The HGAC met for the first time on 27th February 1997, a few days after the announcement of Dolly's birth, to consider its work plan and its initial priorities. With regard to "scientific progress at the frontiers of genetic research which have implications for human genetics" the HGAC (HGAC, P/97/180) proposed to hold a consultative meeting with scientists on the issue. It also decided to contact the HFEA on the adequacy of the existing law to prohibit cloning (ibid.). The HGAC affirmed "its commitment to listening and making itself open to wider public views on the issues at stake" (ibid.).

In September 1997, the HGAC established a joint cloning working group (CWG) together with the HFEA to take forward a public consultation exercise on human cloning as a response to problematisation of public concerns. The consultation was to help in advising the administration on how to deal with 'human cloning' in face of public concerns. The working group consisted of four members, two from the HFEA and two from the HGAC, three of which were either from the life sciences or associated with the pharmaceutical industry. The members for the HFEA were Dr Anne Mc Laren, Wellcome Trust/Cancer Research Campaign Institute, Cambridge, and Prof Christine Gosden, Professor for Medical Genetics at the University of Liverpool; and for the HGAC Rev Dr John

41 Cf. HGAC webpage; at http://www.advisorybodies.doh.gov.uk/hgac/index.html; Oct 20, 2007.

42 At that time, the HGC also swallowed two more advisory bodies, the Advisory Committee on Genetic Testing and the Advisory Group on Scientific Advances in Genetics.

Polkinghorne, Chairman of the Advisory Committee on Genetic Testing, and Dr George Poste, Chief Science and Technology Officer for SmithKline Plc.. Polkinghorne chaired the working group.[43] The CWG issued their consultation paper *Cloning Issues in Reproduction, Science and Medicine* in January 1998 (HFEA/HGAC 1998a), sending it out to numerous, mainly science-based, organisations and individuals, as well as to the members of the Genetics Interest Group (GIG).[44] The consultation was also announced in newspapers and available via the internet in order to reach the general public to at least some extent. Almost 200 organisations and individuals responded to the consultation, including scientific or ethics professionals or organisations, church representatives, and patient groups.[45]

Public Mistrust and Public Misinformation

Again public mistrust and misinformation about science and technology were portrayed as key problems demanding a response—this time through public consultation. However, although called a public consultation, the exercise itself tells a different story, as is also the case with the assertions made by the responsible HFEA/HGAC members. In an article published in spring 1998, Ruth Deech (then chair of the HFEA) and Colin Campbell (then chair of the HGAC) described the motives and goals of the consultation as follows: "it is necessary to clear away the hype surrounding cloning and replace it with informed opinion" (Campbell/Deech 1998). As Deech and Campbell wrote, commentary on the Dolly technique had been over-

43 In the wake of establishing the Donaldson group several years later (cf. 3.7), the British government's consultation procedure was accused of having been influenced by the biotechnological industry because three of the CWG members came from the biotechnological sciences, two of whom were associated with SmithKline and Wellcome (Daily Telegraph, Apr 3, 2000).

44 As Anne Kerr (2003, 213) maintains, the Genetic Interest Group is far from simply being a grassroots organisation for patients and works closely together with the medical and scientific establishment. It supports genetic research and the expansion of genetic services.

45 Although the consultation report announced that responses would be published on the internet (HFEA/HGAC 1998b, para. 2.6), due to the transfer of the issue from the Department of Trade and Industry to the Department of Health, the intention to transfer the responses to electronic format was never carried out (personal email of the Department of Health). As there exists no publicly available list of respondents, I have sifted through the responses in the Department of Health, London, where they are kept today.

shadowed by "fears that this technique might open Pandora's box" and that the consultation should help to avoid a "knee-jerk reaction" (ibid., 31). Already from the outset, public mistrust was associated with misinformation, and being engaged with public concerns was presented as providing 'correct' knowledge and hence as public education. The exercise was supposed to clarify the importance for "all of us" to become more knowledgeable about the subject so as to improve informed debate and to draw attention to "what the consultation paper calls 'therapeutic' cloning types of technique" (ibid.). Already on commencing the consultation process, public opinion, although presented as a serious problem, obviously only played a very limited role: it was only "*envisaged* that this [consultation; slh] paper *will be revised in the light of comments received*, and forms the basis of advice from the HGAC and HFEA to Ministers" (HFEA/HGAC 1998a, 1, emphasis added). This also explains why large parts of the consultation paper and final report were identical or at least very similar. Additionally, the actual impact of the responses on the final report was, at the very least, questionable. For example, although mentioned in the final report, the fact that many respondents disapproved of *any* type of embryo research—including the new cloning research—was dismissed as outside the scope of the consultation. The 'old question' of embryo research was not to be reopened, so the report stated, as it had been discussed at length in the 1990s debates of the HFE Bill. Through this move, many critical responses could be sieved out, although respondents were aware of the difference between so-called 'old' and 'new' questions.

The design of the consultation paper itself mirrored the fact that public mistrust was closely associated with misinformation. Large sections of explanations of the technical process as well as of the different aspects identified by the CWG as important resembled a high-school exam rather than a mature consultation exercise. All sections ended with questions that were not among those to be answered in the response, but which instead functioned as 'guideposts for discussion', reminiscent of school textbooks. Likewise, the actual consultation questions (HFEA/HGAC 1998a, Section 9) resembled 'exam questions' rather than qualified consultation questions. For example, they always referred to the related explanatory paragraphs of the paper and hence appeared as if they were testing whether or not readers understood the paper's content. Of course, respondents could abstain from the suggested format, and many did. However, the responses generally mirrored path dependency in terms of the framing of the problem in the consultation. Most respondents followed the chain of questions without diversion. Nonetheless, the working group was apparently dissatisfied with the 'misinformation' of respondents. As they stated in their report, "responses received to the consultation indicated the need for more education and informed debate" (HFEA/HGAC 1998b, 20).

The questions raised in the paper were generally very technical and scientific. This led many (non-scientific) respondents to complain that they were unable to answer them. Questions such as "Are there any medical or scientific areas that might benefit from research involving human nuclear replacement?" (HFEA/HGAC 1998a, Section 9, Q2) were clearly directed at scientists working in the field. Some respondents criticised that the questions were directive because they were based on the speculative assertion of potential benefits or that socio-cultural questions were lacking altogether. Especially asking for suggestions on "ways to build public confidence in and understanding of the new developments" (ibid., Q9) led some respondents to claim it was assumed that measures had to be taken to alleviate public misunderstanding of the developments rather than to mitigate public disapproval. The educational rather than consultative impetus of the exercise was likewise displayed by insisting on the necessity to reveal the still 'overshadowed' benefits. This stance builds on the belief that, firstly, public disapproval was based on misunderstanding developments rather than on 'serious concerns' and that, secondly, the so-called benefits were beyond scrutiny and without alternatives. Consequently, the final report "takes account not only of the responses to the consultation, but also more recent scientific reports in this fast moving field" (HFEA/HGAC 1998b, foreword, 1).

'Therapeutic' versus Reproductive Cloning

With regard to problematisations in the CWG endeavour, one could observe a transformation of the issues of concern as compared to the STC examination. Along with Dolly, the consultation paper asserted that concerns had been voiced about "the future evolution of this technology, particularly in the context of cloning human beings" (HFEA/HGAC 1998a, 1). Whereas the consultation was directed at "the various ethical issues involved in 'therapeutic' and reproductive cloning", the concern of the CWG was "'human reproductive cloning', which would involve the creation of a human being who was genetically identical to another" (ibid.,

3). In the consultation process, the problem was framed as one of weighing up beneficial and risky, that is, 'therapeutic' and reproductive, applications of human cloning. While the STC had included the management of 'public concerns' on the agenda, the HFEA/HGAC framed these concerns as 'concerns about risks'. These risks were codified as 'reproductive cloning' as opposed to 'benefits' codified as 'therapeutic' cloning. Indeed, the HFEA/HGAC consultation paper was the first official source that introduced the terms 'therapeutic' and 'reproductive' cloning into the debates and emphasised a strict distinction between the two:

"This consultation paper has identified different potential uses of cloning technologies, as this will help to identify the various ethical issues involved. For the purpose of this consultation we draw the distinction between two types of cloning: on the one hand, human reproductive cloning, where the intention is to produce identical foetuses or babies; and, on the other hand, what may broadly be called therapeutic cloning, which (although not coterminous with conventional scientific usage) included other scientific and medical applications of nuclear replacement technology" (ibid., 1).

Different cloning techniques and applications were described at length in the paper (ibid., sections 3 & 4). Additionally, the authors identified the "ethical implications" of "potential research and therapeutic benefits" (ibid., section 7) and of "human reproductive cloning" (ibid., section 8). Under the heading of "*ethical* implications", the paper described several potential outcomes of the research, the so-called "therapeutic benefits", such as the improvement of knowledge about the origins of cancer or ageing (ibid., 8), research on inhibiting mitochondrial disease transmission (ibid.), or generating in-vitro stem cells for gaining insight into how to stimulate regeneration of damaged tissue (ibid., 9). It thereby hinted at the fact that these applications "would not involve the production of a genetically identical individual or fetus" (ibid., 8). With regard to human reproductive cloning, the paper identified the following "ethical implications": the production of genetically identical human beings (ibid., 9); assistance in human reproduction (ibid., 11); safety issues, such as the wastage rates and uncertainty about malformations resulting from the Dolly technique, uncertainty about questions of age[46] (ibid.), or the treatment of human beings as a mere means to an end. The paper also raised questions about the 'eth-

46 'Questions of age' concern the 'difficulty' to specify the age of a clone: How old is Dolly six months after her birth? Six months or six months plus the age of the sheep from which the mammary cell was taken?

ical implications' of the closeness of the technology to 'natural' processes, i.e. 'artificial reproduction' of the "natural process by which monozygotic (identical) twins are produced" (ibid., 9) via embryo splitting, leading to the production of two individuals that share a "genetic identity". Occasionally it calls identical twins "naturally occurring clones". In the light of these 'thoughts' the consultation asked: "To what extent can a person be said to have a right to an individual genetic identity?" (ibid., Q3, 12).

Several respondents not only questioned the concept of the *right to an individual genetic make up* but furthermore doubted the appropriateness of the question altogether and claimed that social and ethical questions concerning cloning should be raised instead. Many commented that the term 'right' was inappropriate in this context. Furthermore, many stated that the paralleling of a 'natural' and an artificial process did not mean that the artificial process was ethically acceptable. The final CWG report, however, stated that "this [inadequacy of the claim to a unique genetic make up; slh] was acknowledged by half the respondents" (HFEA/HGAC 1998b, 15). While some respondents clearly pointed out the principal inadequacy of the concept of 'rights' in this issue area, the working group asserted that "there is no right to a personal genetic make up" (ibid.). Consequently, the final report concluded that there are no ethically relevant issues arising in regard to the protection of genetic identity. As the final report maintains:

> "the existence of a clone […] is not in itself a threat to an individual's identity. Therefore, the claim that each person is entitled to a unique genetic make-up is a correspondingly questionable assertion. Of itself it could not prove an adequate ethical objection to human reproductive cloning" (ibid., 15).

Thus, the consultation paper and the report not only framed the problem as balancing the risks and benefits, but also gave priority to particular interpretations of 'benefits and risks' and provided frames for interpretation and response. The CWG identified the problem, invited participation and response, set the frame for responses, and interpreted the responses in their own terms even if respondents articulated obviously disparate views on the matter.

Maximising Potential and Minimising Concerns

In their final report the CWG described their goal as follows: "it is our wish that the wide potential benefits of this technology are maximised, while at the same time concerns are recognised and adequate safeguards are implemented" (HFEA/HGAC 1998b, 1). In December 1998, the cloning working group recommended that existing legislation be amended to provide for two more research objectives for embryo research: research into therapies for tissue and organ illnesses, and research into mitochondrial diseases, thereby implicitly allowing 'therapeutic' cloning and embryonic stem cell research. The report also stated that the HFE Act did not prohibit cloning and queried whether or not the HFEA had any authority over cloned embryos, because they were not the result of fertilization (the definition of an embryo in the HFE Act) and thus not necessarily subject to HFEA licences.[47] In essence, consulting the public did not make a great impact on the final report. As was already mentioned, many sections of the final report and the consultation paper itself were similar even literally. Although public mistrust was represented as 'the problem', the actual voice of those consulted was made to be inaudible. Public concerns that were equated with misinformation were therefore addressed by recommending more scientific education. The exercise of a public consultation process, so can be concluded, was an act of discourse stimulation directed at securing the maximisation of 'potential benefits of the technology' and inviting public participation in a preframed, educational debate in which technoscientific development was the undisputed prerequisite.

The British government, again, did not accept the recommendations made in the report. Instead it stated in its response that further consideration was needed before they can make decisions (DoH 1999). One reason for its hesitation might have been the opposition to GM crops within the British public at that time. The government may have feared that such scepticism could spill over into human genetics, meaning that public support for the new regulations was an open question (cf. Sexton 2000). The cloning issue was therefore handed over to the Chief Medical Officer, Liam Donaldson; a move which many interpreted as a delaying tactic (ibid.).

47 The HFEA's jurisdiction over cloned embryos was subsequently challenged in court by the Pro-Life Alliance on the grounds that a cloned embryo is not an embryo under UK law because fertilization is not involved. The court ruled that the HFEA had no authority over cloned embryos. The UK government then rushed through legislation at the end of 2001, prohibiting reproductive cloning (The Guardian 2001). The court's ruling, however, was subsequently overturned on appeal.

Summary: Consulting the Public?

The HFEA/HGAC consultation of the public on the question of cloning was again guided by a twofold deficit model: a deficit of information and deficit of trust. In this respect it had an educational impetus, claiming that 'knowledge' was the prerequisite for a rational discussion of the issue of cloning and necessary in order to avoid 'fears' and 'knee-jerk reactions'. The consultation should therefore, at the very least, serve to 'improve informed debate'. Besides the educational impetus concerning the content (information), there is also a tendency to implicitly educate on how to deal with the respective issues: the text book/questionnaire style adopted for the consultation paper not only guided respondents in terms of which information to use, but also suggested how to use it, as well as directed the choice of which issue to deal with. By dismissing discussions of the 'old' question of embryo research as being outside the scope of the consultation, the CWG prescribed a certain attitude towards embryo research to be acknowledged as an 'adequate' response: responses had to generally agree to embryo research and focus on the *new* question of 'cloning' instead. Thus, not only a certain level of knowledgeability but also a specific attitude was demanded of respondents in order to be acknowledged as 'true' participants in the consultation. Also the focus on balancing 'risks' and 'benefits', i.e. reproductive and 'therapeutic' cloning, points in this direction: how to think and talk about the issue of 'cloning' is predetermined in the style of the consultation; the "framework for engagement" (Irwin 2006, 316) is acceptance of embryo research as well as acknowledgement of existing, indisputable (potential) benefits. Additionally, respondents had to implicitly accept the mode of engagement: that is, the framing of the cloning issue as an issue of balancing benefits and risks.

Underlying the consultation process altogether were the assumed inevitability of techno-scientific developments as well as the prospect of potential benefits of the technology and their desirability. Both aspects were interpreted as being beyond problematisation and needed to be communicated *to* the public, engaging it in the debate. It was the goal of the CWG to *maximise* the potential of the technology, and there was not even a reference in the consultation to the *possibility* of refusing 'therapeutic' cloning. As one respondent noted in his letter to the CWG: "Nuclear transfer appears to be a scientific technique looking for an application".

3.7 A New Problem: Britain is Seriously Ill or 'Medical Progress with Responsibility'

"Nuclear transfer appears to be a scientific technique looking for an application." Respondent (HFEA/HGAC 1998a)

In June 1999, the British government announced the establishment of an Expert Advisory Group on 'therapeutic' cloning in humans, chaired by the Chief Medical Officer (CMO), Liam Donaldson. Its task was to assess the "extent to which there is an identified need for and interest in research on human embryos and involving cloning" (CMO letter 1999). On 18th August 1999, Donaldson announced the membership and the terms of reference of the group. Again, all members came from expert circles, many of them were geneticists or medical professors at universities, one was a professional ethicist, one a reader in law, and one a Church representative[48] (DoH 2000a, Annex A). The report was presented to the government in August 2000 and the government accepted the recommendations in full (DoH 2000b). Although the report, which was published during parliament's summer recess, supported the recommendations already made by the HFEA/HGAC group, it sought to expand the purposes for which research may be permitted.

Extensions Wanted

Looking at the group's terms of references, it seems clear that, even before they took up their work, the goal of the exercise was to extend the purposes for which embryo research may be permitted under the HFE Act. As the title of the report *Stem Cell Research: Medical Progress with Responsibility* (DoH 2000a) reveals, the starting point of the exercise was technological development and its possible applications. The issues to be dealt with were, firstly, to assess the degree to which embryonic stem cell research was performed in Britain; secondly, to assess the "anticipated benefits" as well as the "potential risks" and "alternative approaches […] to achieve the same benefits"; thirdly, to assess new ethical or social implications of this research; as well as, fourthly, whether or not regulations needed to be extended in order to include this kind of research (ibid., Annex A). With

48 John Polkinghorne, who had also been a member of the joint cloning working group of the HFEA and the HGAC in 1998.

regard to its terms of references, the CMO group was dealing with the problem of if regulations needed to be extended in order to include the new research. The focus of the consultation process was: the technology is there, what can we do with it?

Change of Problem: Britain's Illnesses

Nonetheless, after an introduction, and thus taking a prominent place in the report, a different problem was presented: It was claimed that there was a significant upsurge of chronic illnesses and a lack of adequate and safe treatments for organ and tissue illnesses in Britain.

> "With the fall in mortality and increased longevity came an upsurge in the occurrence of chronic diseases such as heart disease, stroke, cancer, diabetes, Alzheimer's disease and arthritis. [...] In addition, particularly amongst the young and the elderly, accidents continue to produce injuries that have both acute and long term effects on health" (ibid., 16).

Illnesses affecting tissue and organs were presented as a serious problem in Britain. Although transplantation medicine developed immensely since the second half of the 20th century, it was still confronted with major difficulties that "limit more extensive application of replacement of human tissue by transplantation or grafting" (ibid., 17), the reprot said. Whereas the technical problems were largely surmounted, the report stated that the problem of rejection of transplanted tissue could only be reduced by prescribing immune suppressive drugs, which have serious side-effects. Also chronic failure of the transplanted organ was described as serious problem.[49] The group claimed, however, that the greatest obstacle to a wider application of transplantation was the shortage of organs and tissues with over 5000 patients on the waiting list (ibid., 17).

Stem cell research and cell nuclear replacement technology were presented as possible and, in fact, as the only solutions to these problems. Stem cells derived from embryos created via cell nuclear replacement (CNR) could, so the report maintained, be influenced so they differentiated into different kinds of cells and therefore could be used to treat degenerative cell diseases and illnesses affecting tissue or organs, possibly avoiding re-

---

49 Interestingly, the side-effects and problems have usually been rather concealed and played down in the context of promoting transplantation medicine, while they are now highlighted in the context of stem cell research.

jection. "Research involving the extraction and use of stem cells raises the prospect of a range of exciting new therapeutic possibilities for the repair of diseased or damaged tissues in the future, which could eventually bring major health benefits" (ibid., 29).

Another declared goal of cell nuclear replacement was to "help women avoid the birth of a child with inherited mitochondrial disease" (ibid., 27). The transmission of defected mitochondrial DNA, the report said, could be avoided by oocyte nucleus transfer: the woman's egg-shell carrying mitochondrial DNA would be de-nucleated and replaced by a donor egg-shell. The nucleus of the mother's egg would be introduced into the donor egg-shell and then fertilised by the father's sperm.[50] At present, the transmission of mitochondrial DNA can only be avoided by using a donor egg (ibid., 27), a scenario that, according to the CMO group, "is not acceptable to many parents because the natural mother is not playing any part in the genetic make up of the resulting child" (ibid., 28).

The presentation of *serious diseases as a severe problem* in Britain was new in the debate. Although the so-called possibilities of research cloning had been mentioned before, they had been rather vaguely defined and portrayed as speculation. The CMO report was the first (official) report that clearly problematised 'serious diseases' as a societal problem in need of a remedy and simultaneously presented embryonic stem cell research as the appropriate answer.

Even though stem cells issuing from other than embryonic sources were mentioned, for example, stem cells retrieved from adult tissue, such as bone marrow, brain, skin or blood, from aborted fetuses, or from umbilical cord blood, they were declared *not* to be an alternative to embryonic stem cells. The effectiveness of their use could, the report asserted, only be evaluated on the basis of prior research on embryonic stem cells. While the report stated it could be "a long term promise" that adult stem cells equal or even surpass the effectiveness of embryonic stem cells (ibid., 19), research on embryonic stem cells was depicted as unavoidable also in the development of adult stem cell research. Other solutions than 'stem cell

50 Oocyte nucleus transfer differs to what is called 'cell nuclear replacement' in the report in that it involves the nucleus of the egg cell and not that of an adult cell. Thus, in contrast to CNR, the resulting cell has to be /can be fertilised. After fertilization the result consists of parts of three, and not two, different cells (egg shell, egg nucleus, sperm). However, both techniques involve the transfer of cell nuclei and are often both referred to as cell nuclear replacement in the literature.

solutions' were not discussed. Thus, the nexus serious diseases-embryonic stem cell research was presupposed and, admittedly, extremely vague. Embryonic stem cell research was presented as a *solution*, leaving the 'need' for embryonic stem cell research beyond problematisation.

As we know from the declaration of tasks, the group was set up to assess the potential benefits arising from stem cell technology and not to assess the problem of serious illnesses. Its target was not to appraise the problem of illness, but to determine the "areas of human health in which the use of therapeutic cloning techniques is most likely to provide benefits" (ibid., 51). Thus, we can interpret problematisation of 'Britain's illness' in the opening paragraph of the report as a mode of rationalizing and legitimizing embryonic stem cell research rather than actually politicising and discussing the problem of *illness* in Britain. The rather sketchy presentation of the problem 'serious diseases' as well as the lack of discussion of alternative solutions makes this very evident. Through this move, embryonic stem cell research was declared to be a possible solution without an alternative.

More Problems and More Uncertainties: The Possible Potential of Science

However, while presented as a solution to the problem of illnesses, stem cell research and CNR harbour problems of their own, which were the focus of attention of the expert group. The performance and development of the new technologies required the donation of thousands of women's egg cells (ibid., 24). The current lack of eggs from "live human donors" (ibid., 25)[51] for fertility treatment implies, so the report said, an obstacle for the development of the cell nuclear replacement technique. However, given the concerns about the use of embryos in research and the shortage in egg cells, a long-term goal of ES cell research was the development of techniques to reprogramme adult stem cells and revert them into an unspecialised state (ibid., 25). Research on embryonic stem cells was portrayed as a major and inevitable step in this direction (ibid., 40).

The most prominent safety issues arising from CNR technology were considered to be risk of tumour formation, because stem cells may carry mutations or defects arising from the reprogramming process and on account of their DNA not matching that of the recipient (donor of cell nu-

51 The report avoids the use of the term *women* throughout the whole text.

cleus) exactly as the egg shell also carries mitochondrial DNA. Nonetheless, CNR technology could, as the report asserted, "prevent the child from inheriting the disease" (ibid., 28). However, "very little research work has been carried out to establish whether this *theoretical* reasoning would be confirmed in experiments and whether the technique would in fact lead to the prevention of disorders of mitochondrial diseases" (ibid., 28, emphasis added).

Another 'risk' discussed in the report was reproductive cloning, that is, the use of cell nuclear replacement in order to establish a pregnancy. This, however, could not take place in the UK, the CMO report stated, as the HFEA had made clear that "it will not issue a licence for treatment involving the use of embryos created this way" (ibid., 34). If this was understood to be insufficient, the group recommended additional controls via a new act of parliament (ibid., 47).

Altogether, as was the case also in the STC report, anticipated research results were depicted in a highly speculative and theoretical way. The extensive use of conditional phrases and terms, especially with regard to prospective treatment possibilities, hints at an enormous uncertainty about the so-called benefits of the research. The following quote illustrates the vagueness of the whole project of stem cell research and CNR:

> "It is *envisaged* that tissues developed from stem cells derived from embryos created by cell nuclear replacement *would* have the advantage of being free from rejection because they *should* be genetically compatible with the person being treated, from whom the donor nucleus was taken. *If this proves to be the case*, there is a great *potential* benefit to human health" (ibid., 23, emphasis added).

The uncertainty of research results, however, underpinned the call for more research rather than the search for alternatives. More research was presented as a solution to the problem of uncertainty of research results. The advisory group predicted that "the scientific and technical hurdles that need to be overcome will require many years of basic research. Before it is possible to use stem cells effectively for therapeutic purposes, fundamental problems in the field of cell and development biology will need to be solved" (ibid., 30). Encouraging further research would, they suggested, "enable a better assessment to be made of the true potential of the research and the most fruitful avenues to pursue further" (ibid., 31). Likewise, the group emphasised in the executive summary of their report that they considered "cellular research and development" to be "basic research which if permitted would precede, probably by many years, any possible application to treatment" (ibid., 5). The main goal of the new research was therefore to understand and evaluate its *possible potential.* Although presented as a solution without alternatives to 'the problem' of serious diseases, the expected results of the research were only speculations. The speculative nature of anticipated goals, however, only supported the call for more research. Thus both levels of problematisation—of 'serious diseases' and that of 'uncertainty of research'—were presented as requiring the same answer, that is, more research.

Balancing Ethics

The 'ethical' implications were a further aspect of the new research to be considered by the Expert Advisory Group. The group insisted on the fact that it was supposed to examine "*new ethical issues* that might arise from the creation and use of embryos for the extraction of stem cells for research into new therapies" (ibid., 12, emphasis added). It was thereby stressed that it did not review "from first principles the ethics of research involving embryos" (ibid.). As a strong regulatory system already existed in the UK, embryo research in general was not to be reconsidered.

Whereas in 1998 the HFEA/HGAC had defined the 'ethical question' as the balancing of benefits and risks of cloning, that is, of 'therapeutic' versus reproductive cloning, the ethical question discussed in the CMO report focused more narrowly on the 'status of the embryo' and its possible use for treatment purposes. The problem discussed in the report was framed as a need to balance the status of the *source of stem cells* against possible benefits, such as possible cures for particular diseases and possible elimination of research uncertainty. Problematisation related to embryonic stem cell research bore the appearance of an ethical conflict requiring a balance between two values: those concerning the embryo, on the one hand, and those concerning possible benefits for human health or the development of the research, on the other. As it was their task to consider only *new* ethical issues connected to the extraction of stem cells or the creation of embryos via CNR, the expert group consequently refused to reconsider the question of embryo research more generally. It was exactly this question, however, that was portrayed as *the* moral controversy.

The moral controversy concerning the embryo, the CMO group argued, had already been discussed and decided upon by parliament in the wake of the HFE Bill in the 1980s and was not open to renegotiation (ibid., 37). Consequently, the report discussed the "ethical implications of an *extension* of the research purposes" (ibid., 38, emphasis added) governed in the HFE Act. On the basis of generally accepting the use of embryos for research, the report took up a clearly utilitarian standpoint of 'the more the better': Given that embryo research had already been permitted in the UK in the context of infertility treatment and congenital diseases, the extension of research purposes to include *more* people and *more* disorders was portrayed as morally *more* justified. "Indeed if there is a greater potential for the research to benefit a wider range of people or treat a wider range of disorders such research may be more ethically justified" (ibid., 39).

Serious consequences of the construction of an 'ethical' problem of a balance between the status of the embryo and possible benefits were not restricted to the conception of ethics and to the range of issues that could possibly be discussed. The construction also had a profound impact on the format of the report. The chapter on ethics was arranged in such way that different moral views on the status of the embryo were weighed against each other, simulating a 'real' debate. However, real actors were not mentioned. Referring to unspecified groups of people, the expert group claimed that the status of the embryo and its use in research were controversial issues. While the group considered the potential of embryonic stem cells to be very promising, especially for the treatment of illnesses affecting tissue and organs, they also claimed that "the source of stem cells [...] raises important ethical issues for many people" (ibid., 13). At one end of the scale, a "significant minority of people believes that the use of any embryos for research purposes is unethical and unacceptable on the grounds that an embryo is a human being entitled to full human status from the moment of its conception and, in particular, has a right to life" (ibid., 37). At the other end, there were those who believe that the embryo is a mere ball of cells, similar to any other collection of human cells, and could thus be used in research (ibid.). The middle position would regard the status of the embryo as a continuum, increasing parallel to the development of the embryo (ibid.). According to this view the report claimed that at the early stage of embryo development it is "morally justified to use embryos for research purposes in order to benefit others, provided that such research is necessary and justified by the benefit it may produce in the long run" (ibid.). The report did not mention who the representatives of the respective views were. In their descriptions of various 'moral views' they used phrases such as *it could be argued* or *a contrary moral view is* without giving references or quoting a particular individual or group of people. Therefore, it is impossible to retrace, scrutinize, or even question these views, and the whole endeavour of an 'ethical debate' seems highly fictitious. Although it was said that the expert group received over one hundred responses (ibid., 12), the report itself only provides a short list of so-called 'key references', which includes science-based articles, a few international declarations or reports, guidelines for research, and legal texts (ibid., 53). The only references stemming from a clearly identified British actor and also the only source obviously considering 'ethical issues' is a report by the UK Nuffield Council on Bioethics (cf. Nuffield Council on Bioethics 2000). Except for 'scientists' in general and 'persons affected with diseases', no actors were specifically mentioned in the report. People suffering "from a wide range of incurable disorders" (DoH 2000a, 37) and researchers were declared to be enthusiastic about the developments and its possible progress. At the other end of the discussion, those associated with the "strongly held views opposed to the use of embryos in this, or indeed any, research" (ibid., 12) remained unspecified. This approach in dealing with the issue corresponded with the construction of the problem as an *ethical* problem of balancing pros and cons and not as one of, for example, conflicting different interests of various actors. 'Ethics' was depicted as a balancing process: as balancing the status of the embryo and possible benefits, on the one hand, and, on the other, as a confrontation of opposing moral views on this balance. However, the group concluded: "It is not possible to reconcile the opposing views on the moral status of the embryo and on the use of embryos in research" (ibid., 39). The ethical discussion within the report ends as follows:

> "Having given careful consideration to the spectrum of ethical opinion, the Expert Group concluded that, as with the currently permitted uses of embryos in research, the potential benefits of the research could be weighed against the respect owed to the embryo given its very early stage of development. The question was, therefore, whether the potential benefit of research involving extracting stem cells from early embryos was sufficient to justify the use of embryos in such research and whether there were alternatives to embryo research offering equivalent benefits" (ibid. , 41).

Law and Control

As it was the task of the expert group to discuss whether or not the law needed to be extended in order to allow stem cell research, it is not surprising that legal matters were discussed at length. The report described the process of the development and implementation of the HFE Act in the 1980s as well as juridical aspects of embryo research, also including the number of embryos created between 1992 and 1998 in addition to how they were used. The report described the purposes for which embryo research was used and already legal in Britain.[52] It emphasised that in this phase the *extraction* of stem cells was permitted under the HFE Act, provided that they are used for one of these purposes. After stem cells have been extracted, however, the HFE Act is no longer applicable for research to be carried out on them.[53] Furthermore, the creation of embryos via cell nuclear replacement, the report continued, was not banned by the act and was ultimately dependent on the legality of the purpose of research. The CMO report asserted that the use of embryos in research for *non*-congenital diseases had been illegal to date. The act would, however, allow further research purposes to be added if both Houses of Parliament agreed, although it restricted the kind of purposes that may be added "to research which would increase knowledge about the creation or development of embryos, or about diseases, or which would enable such knowledge to be applied" (ibid., 34). The legal problem following the success of this type of research would be whether or not embryos might be used for so-called treatment purposes, such as the creation of tissue via cell nuclear replacement technology (ibid.). This would have to be considered by parliament. The report claimed that cell nuclear replacement for subsequent implantation into a woman's womb to establish a pregnancy was impossible in the UK because the HFEA would not grant a licence for treatment involving the use of embryos created in this way (ibid., 34).

The Chief Medical Officer's Expert Advisory Group recommended that research purposes be extended in the HFE Act to include "research using embryos (whether created by in vitro fertilization or cell nuclear replacement) to increase understanding about human diseases and disorders and their cell-based treatments" (ibid., 45). Such research should be subject to the controls established in the HFE Act. In licensing this research the HFEA should be satisfied that it is not possible to meet the goals of the research in alternative way (ibid., 46). The group also demanded that individuals donating eggs or sperm should specifically consent to the use of the resulting embryos in stem cell research. The new purposes

52 The foregoing legal purposes for embryo research are described in chapter 3.2.
53 The HFE Act only applies to the use of embryos and not to that of embryonic cells.

should include research to increase understanding of mitochondrial diseases and their treatment using cell nuclear replacement technology, again subject to the controls stipulated in the HFE Act. The HFEA, together with the Human Genetics Commission,[54] should monitor research involving stem cells and constantly review information on the developments made therein as a basis for advising ministers on progress in human ES cell research, on the implications for healthcare, as well as on unforeseeable concerns that could arise. The creation of human-animal hybrids should be forbidden and the transfer of an embryo created by cell nuclear replacement into the uterus of a woman (so-called reproductive cloning) should remain a criminal offence. If necessary, this should be strengthened by a new act of parliament. In case the recommendations were implemented and in case the research was successful, the Expert Group suggested that they keep the possible need for legislation allowing the use of embryonic stem cells in medical treatment under review. Last but not least, the Research Council should support the establishment of a programme for stem cell research and consider the feasibility of establishing collections of stem cells for research use. The council should also be encouraged to "give high priority to this research" (ibid., 48).

The Government Response

In August 2000, the British government fully accepted the recommendations made in the Chief Medical Officer's report and confirmed to bring legislation forward where necessary, specifically stating that it would unequivocally implement the ban on reproductive cloning. In its response (DoH 2000b) to the CMO report, the government also declared its intention to allow a free vote in parliament on the recommendations made therein. Regarding the recommendations referring to the new research purposes of human embryo research and the possible 'ethical' reservations against them, the government stressed that it "respects the view of those people who believe that human embryos should not be used in any research" (ibid., 4). Nonetheless, it considered the potential benefits of the new research to justify the new research purposes. Thus, the view that research should not be carried out with human embryos was outweighed. 'Progress' in research should, the government went on, be kept under

54 The successor of the Human Genetics Advisory Commission.

review with regard to the need of new legislation on, for example, the use of cell-based treatments or other developments not yet known (ibid., 6). The government also accepted the recommendation to establish a research programme on the derivation of stem cell lines from all sources, on the production of stem cell lines via CNR, on reprogramming somatic cells, as well as on stem cell differentiation for treatment purposes.

Summary: We've Found a Problem to a Solution

The most significant contribution of the Chief Medical Officer's Expert Advisory Group consultation and report was to introduce a new problematisation into the debate, namely the problem of 'serious diseases'. By making this move, it transformed research cloning into a *solution* without alternatives, and 'problems' with stem cell research were deflated into second order problems. The second order problems were subdivided into technical problems and ethical concerns, whereby, following Carr and Levidow (1997), the former applied to the domain of objective science and expertise, and the latter to the realm of 'irreconcilable moral beliefs'. In this way, 'ethical' problems, such as concerns or unease, could be addressed without, however, seriously challenging or having serious impacts on the 'technical' side. Indeed, 'ethical' problems were depicted as irreconcilable moral views, which, as the government asserted, 'are respected' but cannot form the basis for science policy-making. 'Sound science' was not replaced or reflected upon, but the (not so) hidden agenda of ongoing scientific development remained unchallenged.

Nevertheless, the 'necessity' and inevitability of scientific development was underpinned by references to 'serious diseases'. Possible cures formed the moral value that had to be balanced against arguments for the moral status of the embryo. The possibility of curing serious diseases justified the call for the extension of research purposes. 'Health' was the social vision that science could provide or at least make possible. Additionally, references to scientific 'problems', i.e. to the uncertainty of research results, also corroborated the call for more research: it was claimed that such uncertainty could be 'overcome' only through more research. Therefore, in view of the 'terms of reference' of the CMO group, it is not surprising that the recommendations made in the report referred to the legal aspects restricting the possibilities of the new research and suggested the extension of research purposes in the HFE Act. The Chief Medical Officer's report served as background information for the parliamentary debates on the issue. The debates, which took place between November 2000 and January 2001, will be discussed in the following chapter.

3.8 A Question of Rights and Individual Moral Responsibility

"Accepting treatment is a personal decision." Evan Harris (HC, Dec 19, 2000, col. 253)

The debates in the British Houses on stem cell research took place between November 2000 and January 2001.[55] They took place in a legal context that already allowed embryo research in the context of fertility treatment and procreation in the UK. Previously, the HFE Act permitted embryo research up to 14 days after fertilisation if it served one out of five research purposes and if the research goals could not be met in other ways (cf. 3.2).[56]

Despite the fact that the government had accepted the recommendations of the CMO expert group, a few days before the House of Commons voted the government published the draft regulations, which exceeded Donaldson's suggestions. While the expert group had suggested that two more research purposes in embryo research be allowed, on the 19th December 2000 the House of Commons voted on three. Like the CMO recommendations, the government's draft regulations would permit embryo research (under licence of the HFEA) if it served: (1) increasing knowledge about serious diseases and (2) enabling such knowledge to be applied in developing treatment for serious diseases. The government's draft added, however, the purpose of (3) increasing knowledge about the development of embryos, and thus opened up a rather broad field of research options.

With regard to the legislative process on research cloning a few formal parliamentary and political characteristics of British politics must be mentioned. First, the British Parliament has unfettered authority to make any law they wish and no other body has the right to ignore its legislation. Secondly, the British political system is majority-governed, meaning that

55 I have discussed the parliamentary debates in more detail in Herrmann (2003).

56 The five allowed research purposes were: (1) to advance IVF; (2) to understand the causes of miscarriage; (3) to improve contraception; (4) to develop methods of detecting genetic or chromosomal deviations in embryos before implantation; and (5) to understand congenital diseases (cf. Morgan/Lee 1991).

the government party can usually implement the policies they wish.[57] Also, the Upper House has restricted legislative power: If a bill has been accepted by the House of Commons, the House of Lords reviews and can reject it. However, if the House of Commons sends the same bill back a year later, it becomes an act even if the House of Lords rejects it again. Thus, in effect, the House of Lords can only delay a bill for one year and cannot—if the House of Commons does not agree—reject it altogether. The new research purposes were presented by Yvette Cooper, then under-secretary of state for health, as secondary legislation, that is, as an extension of existing legislation, rather than primary legislation. By doing so, parliamentarians could not amend the draft regulations but only approve or reject them. This fact has been subject to some dispute during the debates in both Houses. Especially opponents to the new purposes criticised this move as misjudging the importance of the changes in law. They claimed that there was not enough time to discuss the regulations and that introducing secondary legislation was a strategic move by the government in order to prevent amendments and more parliamentary debates. The government argued that there had been enough debate, especially as there were no *new* ethical questions since 1990. The extension of research purposes in the HFE Act, however, was subject to a free vote.

Despite these differences and despite the fact that parliament could not amend the regulations, the MPs agreed that the ongoing debates were exceptional and exemplary for parliamentary debates in general. Many commented on the high quality of the debates, although most MPs declared that they would not change their minds on the issue.

Nothing New: Embryos and the Greater Good

The introduction of secondary legislation to extend the research purposes was justified in the debates by stating that the new research was "in keeping with the HFE Act" (Baroness Northover, HL, Jan 22, 2001, col. 35). The new purposes were presented by supporters as 'nothing new', as minimal changes to the law not entailing any new ethical or social implications. MPs supporting the changes in law underpinned such an argument by discursively separating 'therapeutic' from reproductive cloning research, referring to a division introduced in the HFEA/HGAC public consultation

57 If the government party loses a case it is usually because of individual party member's opposition.

(cf. 3.6) as well as in the Donaldson report (cf. 3.7) discussed above. They additionally called for a ban on reproductive cloning. Research cloning was defended by supporters on the grounds of a division of the acceptable—'therapeutic'—and the unacceptable—reproductive—application of the new technology. Although the new regulations (or their draft) did not explicitly mention the term 'therapeutic' cloning, the differentiation between 'therapeutic' and reproductive cloning formed an important argument in the debates. MPs agreed that reproductive cloning should be banned, whereas supporters claimed that the so-called 'therapeutic' application of the technique should be allowed. Because of its potential for developing cures for 'serious diseases', it was declared to be hardly morally questionable but even morally required. At the same time, supporters managed to portray the new research as if it were not new but a logical extension of already permitted practices. They stressed that it was not a fundamentally new practice by maintaining that 'therapeutic' cloning and embryonic stem cell research were simply a minimal extension of what already was legal. Therefore, the call for a ban on reproductive cloning was a device for legitimating 'therapeutic' cloning. Opponents countered the separation and confrontation of 'therapeutic' and reproductive cloning as well as the eventual legitimation of the former via the delegitimation of the latter. They feared that 'therapeutic' cloning was the first step in or the same procedure as reproductive cloning. For opponents, cloning was just cloning.

The debates in the Houses particularly focused on the possibility of developing treatments for 'serious diseases', and one of the central topics was the status of the embryo. Already in the debates of the 1980s (on the HFE Bill), the embryo was the focus of attention. Mulkay (1997) observed that every statement appearing in those debates was introduced by an assertion concerning the 'status of the embryo'. Also in recent debates, supporters as well as opponents of the new research purposes concentrated on the embryo, its status, and its right to life or not having this right: While supporters appealed to the embryo's 'gradual status', which is likewise the basic assumption in the HFE Act,[58] others used the debates to question permitting

58 The HFE Act allows embryo research for up to fourteen days. The 14-day rule is based in the development of the so-called primitive streak, one of the first signs of the spinal cord, which is said to appear about 14 days after fertilisation. The Warnock report had introduced this notion in order to prescribe an absolute limitation. This differentiation was strengthened when the notion of the 'pre-embryo' was introduced. The 'pre-

embryo research altogether, basing their arguments on the 'absolute status of the embryo'. While supporters insisted on the fact that embryo research was already allowed and that it was legitimate to extend research purposes, others argued that the embryo was in need of protection and that the change of law would mean a deterioration of the present legal situation, which maintains a 'special status of the embryo'. Others were opposed to embryo research altogether on the basis of the embryo's right to life. As one MP said: "There are those who believe that embryonic cells should not be used for experimentation or treatment. They believe that it is a moral issue purely about respecting the right to life. That is the group into which I would fall" (Dr. Liam Fox, HC, Dec 19, 2000, col. 226).

The legal context of allowing embryo research for infertility treatment or procreation made it easier for pro-research MPs to claim that the new research was not in fact new. MPs argued that research into curing serious diseases was only a logical extension of legally permitted research. This line of argumentation helped to establish an understanding of stem cell research and cloning that did not appear to entail any new ethical, legal, or social implications but only those 'already covered by legislation' (cf. also Parry 2003).[59]

"I do not believe that there is any great new ethical issue here. The issue before us is not whether we should conduct research on embryos. As the minister said, that issue was debated at length during the passage of the Human Fertilisation and Embryology Act 1990. That legislation already allows research on embryos for certain medical conditions such as congenital diseases or genetic abnormalities. At the time of the 1990 Act, no one envisaged the possibility of solutions through therapeutic cloning for say Parkinson's—certainly a disease, and certainly abnormal. If research is acceptable for promoting knowledge about the causes of congenital diseases or advances in the treatment of infertility, why should we not widen the rules to include promotion of advances in the understanding of other human diseases and disorders and to allow the possibility of research to develop new cell-based treatments" (Dr. Peter Brand, HC, Nov 17, 2000, col. 1196).

---

embryo' was one of the most controversial notions in the 1980s debates (cf. Franklin 1999b; Spallone 1989).

59 Sarah Parry (2003) claimed that one of the reasons for the success of this argumentation was the lack of a discussion on the social implications of cloning, which had been impeded by concentrating on the 'status of the early embryo'. However, I would suggest that it was not the *absence* of discussion on social implications but the focus on particular social implications, namely societal 'progress' or curing individual suffering through scientific experimentation that underpinned the 'nothing new' argument.

The comparison of allowed research into reproduction and infertility and proposed research into curing serious diseases provided a strong moral-laden argument for permitting the new research purposes: "if one accepts that basis [the HFE Act; slh] and the fact that it is acceptable to use embryo research—and indeed embryos in fertility treatment—logically one must accept the extension to the potential treatment of all those serious diseases" (Nick Harvey, HC, Dec 19, 2000, col. 230).

Opponents of the new research, however, disagreed that 'all' questions had been discussed in the 1980s: as Ann Winterton (HC, Nov 17, 2000 col. 1175) said, "I cannot recall a debate on cloning" having taken place at that time. Another MP stated that the fact the House of Commons had accepted the use of embryos in research "does not mean that I have to accept it" (Edward Leigh, HC, Dec 15, 2000, col. 889).

Still, the focus of debate on the 'status of the embryo' helped supporters to claim that all 'ethical' questions concerning embryo research had been solved in the older debates and were not open to renegotiation. On the other hand, most opponents generally opposed embryo research and declared that the embryo had to be protected. Their main argument was countered by supporters claiming that it had been superseded by existing law. Hence, opposition to the new research purposes did not seem legitimate or, at the very least, appeared illogical. The framing of 'nothing new' therefore compelled opponents of the new research purposes legalisation to frame their arguments in terms of either a moral or scientific confrontation of adult and embryonic stem cell research or in terms of a general opposition to embryo research. The latter was much more difficult to put forward as it was considered to question the HFE Act altogether.

However, as some 'unease' towards the new research purposes obviously persisted, (also supporters of the new research emphasised that embryo research had to be strictly controlled, e.g. Yvette Cooper, HC, Dec 15, 2000, col. 881), the need for respect for the embryo[60] was set against another central element of the debate—'the greater good (for society)'. [61] Thus, the central problematisation of the debates was a *conflict between 'the status of the embryo' and 'the greater good (for society)'*. The integration of embryo donation and research into a wider concept of a 'greater good' was recur-

60 Some 'unease' is also demanded in the HFEA Code of Practice 1991, which prescribes "the respect which is due to human life in all stages in its development", cf. 3.2.

61 This 'greater good' appeared in various linguistic forms, such as the 'common good', 'greater good for society', 'benefits to society', or 'better good of humanity'.

ring throughout the parliamentary debates. 'Benefits to society', 'the common good', or 'a greater good' were portrayed by supporters as the ultimate goal of 'therapeutic' cloning and stem cell research. It was claimed that the new research, dedicated to the relief of suffering, would benefit both the individual and society, and would therefore lead to overall societal 'progress' through scientific 'progress'. To relieve suffering, as Labour MP Anne Begg suggested, was to "cut waste", since "waste in personal terms and in life is a cost to society" (HC, Nov 15, 2000, col. 911).

Although it was essential for supporters of the new research to portray the legal changes as 'minimal', it was equally important to stress the great significance of embryonic stem cell research and 'therapeutic' cloning for the cause of a proclaimed moral duty to help the sick. The potential of developing cures and the responsibility towards the 'seriously ill' was an important moral—or moralising—factor in supporters' line of argumentation. Individual life stories of people suffering from Parkinson's or Alzheimer's, of children whose parents were close to death, or who were themselves suffering from 'serious diseases', as well as quotes from letters of organisations or individuals served the construction of a (public) *need* for new therapeutic opportunities. 'Serious diseases' were constructed as a great and pressing problem that could only be solved scientifically. The construction of a group of 'sufferers' helped to awaken public interest in new therapies. Thereby a moral duty was generated in parliament to facilitate their development.[62] The discourse on needs was translated into a discourse on rights and, finally, into a discourse on the duty of the state to help the sick: "We should consider for a moment the very real suffering that could be in our power to end by carrying out possible ground-breaking research" (Yvette Cooper, HC, Nov 17, 2000, col. 1181). Or as Michael Fabricant said: "If we use the argument that we should not go ahead and begin the research, we would not be doing justice to the people whom we represent in parliament" (HC, Dec 15, 2000, col. 920).

In contrast, opponents stated that "the most important thing that we can do in this place is nurture human life and resist the materialist and consumerist view of society that human beings, however innocent and however small, are expendable. We cannot accept the argument of the greater good" (Edward Leigh, ibid., col. 888) Still the *greater good for society*

62 I emphasise that I do not deny the existence of suffering or serious illnesses. However, I wish to highlight the function of the discursive figure of 'serious diseases' to underpin calls for the legalisation of research cloning.

was ultimately able to outweigh any 'respect for the embryo' or the recognition of the "sacred nature of human life", serving the interests of the supporters of the new legislation:

"Most members of the public, whether they are religious or would describe themselves as humanist or secular, recognise the special and sacred nature of human life. They believe that there must be respect for that entity. The only circumstances in which research on this cluster of cells should be permissible is when a genuine greater good for society will be served" (Baroness Kennedy of The Shaws, HL, Jan 22, 2001, col. 47).

Nonetheless, the discursive construction of a group of sufferers in need of help and the moral duty of parliament to provide help was based on a discourse of hope and great uncertainty.

"None of us knows exactly what discoveries stem cell research could lead us to—scientists do not know, ethicists do not know, and parliamentarians do not know—but potentially revolutionary treatments lie within our grasp if this House and the other place give the regulations the go-ahead. All hon. Members must weigh up the complex arguments for themselves, but we must be aware of the power that we could be holding in our hands as we consider these issues" (Yvette Cooper, HC, Nov 17, 2000, col. 1182).

As was the case in the consultations and statements discussed in the previous chapters, also MPs were highly uncertain about research results, and conditional modes of speech predominated in the contributions with regard to what scientific experiments could actually achieve. This did not, however, hinder the demand for the expansion of scientific research but, on the contrary and as far as the supporters were concerned, fostered it. Indeed, they maintained that it was precisely the unpredictability of results that made research even more desirable. As Labour MP Anne Begg stated: "It is precisely because scientists do not know the answer that we need the research. If research does not work, that is fine; nothing else will happen. However, we must have the research in the first place" (HC, Dec 15, 2000, col. 902).

Open to various interpretations, the discursive figure of 'serious diseases' very comfortably qualified arguments allowing for belief in strict limits and control of the new purposes, while anticipating extended boundaries and the inclusion of more and more diseases rated as 'serious' as scientific research proceeds. Even though this figure is strongly moral-laden it nonetheless does not require explanation. Indeed, a definition would be rather difficult to legitimate.[63] The notion of 'serious diseases' can easily sanction scientific or medical developments, whereas it is difficult or impossible to argue against finding a cure for them. And indeed, at least in the stem cell debate, not only the classical 'serious diseases' Alzheimer's or Parkinson's were mentioned but also burns, strokes, or other injuries or illnesses. However, as was also the case in the Donaldson report introducing problematisation of 'serious diseases', alternative responses to this problem—except for research on adult stem cells that opponents put forth—were not put forward. This speaks for problematisation that supports a certain solution—namely stem cell research—rather than searching for a solution to a specific problem.

Science, Science, Science

A further framing that appeared in the parliamentary debates was the inevitability of scientific development, an argument employed by both supporters and opponents of the new regulation. Supporters defined scientific research and 'progress' as generally benefiting society even if damaging effects were also involved. As conservative MP Michael Fabricant put it:

> "we should never use the argument that we can never undertake any experimentation for the better good of humanity if it might lead to research that could damage society […] where would that lead? There would never be a change in society. […] God has given us an intellect, which I believe should be used for the greater good" (Michael Fabricant, HC, 15 Dec. 2000, col. 919).

Science was defined as a practice that, in general, supports the 'greater good' and improves society[64] as a means of providing necessary information or knowledge "which allows us to lead our lives and create society in a better way" (Dr. Peter Brand, HC, Nov 17, 2000, col. 1196). With this, the restriction of scientific knowledge production was framed in terms of risking societal decay. Also the opponents of the new regulations defended

---

63 As the experience with pre-implantation genetic diagnosis in the UK shows, it is in fact very difficult to keep such borders intact. The HFEA, which licenses the checking of embryos in-vitro, gradually extends the list of diseases they grant licences for. In fact, it would be rather difficult to justify drawing up a hierarchical list of illnesses according to some kind of 'standard of seriousness'.

64 Cf. Sarah Franklin's (1997, 99) discussion of "the vision of natural science as a means of implementing social progress through the redesign of reproductive processes", which, according to the author, is not a recent innovation in Great Britain.

the 'need' for scientific 'progress', arguing, for their part, that prohibition of 'immoral' research would not hinder 'progress'. Germany served as an example for the conservative MP Ann Winterton, who stated that although Germany had strict regulations regarding embryo research in Europe "its biotechnology industry has outstripped the industry in the United Kingdom" (ibid., col. 1205).

At the same time scientific 'progress' was portrayed as an inevitable, as good as 'natural' development: "We must accept that science is a continuum" (Michael Clark, HC, Dec 19, 2000, col. 249) in which Britain should participate. Firstly, because it always did, as Labour MP Ian Gibson stated: "in the scientific and medical field, other countries realise that there is much work to be done and much knowledge to be gained. We cannot sit back and allow that to go on in the world of international science and medicine, without playing the dominant role that we have had for many years" (HC, Dec 15, 2000, col. 892). And secondly, because participation in research would allow participation in exploiting the benefits that arise from the results: "it would also be moral hypocrisy to allow the research to be done abroad and then use its findings in this country after we had failed to pass the regulations" (Dr. Michael Clark, HC, Dec 19, 2000, col. 245).

Also some opponents of the new research shared the view that science will inevitably progress somewhere in the world: "The issue is not for or against stem cell research. […] The research will continue, whatever we decide" (Philip Hammond, HC, Nov 17, 2000, col. 1187). The question was, however, "whether we, within the limited scope of our jurisdiction, wish to sanction such activity" (ibid., col. 1188).

Against the background of the belief in the inevitability of scientific progress, at a political level we find a strong degree of resignation towards the option of restricting scientific development. This can be especially detected in supporters' arguments, but it is also prevalent in those of the opponents to the new regulation. Most MPs seem to agree to an imperative of scientific progress in face of international science and research, or as Yvette Cooper, maintained: "We have in our power to authorise the research" (HC Dec 15, 2000, col. 880), but not necessarily the power to authorise its restriction. This is in line with the fact the reports and statements discussed above usually stressed that regulation of science must be flexible in order to be able to adapt to scientific advances.

Science, Ethics, Politics?

Apart from the imperative of scientific progress, which would make any debate on regulation redundant, most MPs expressed that there were also some 'ethical' issues related to scientific developments that necessitated debate and demanded caution and control. Also the straightforward supporters of the new research purposes emphasised that "the research will proceed in this country only under closely regulated conditions" (Baroness Greengross, HL Jan 22, 2001, col. 86). Additionally, the strict separation of reproductive and 'therapeutic' cloning in supporters' arguments served as a discursive means of defining the borders between illegitimate and legitimate techno-scientific practices. However, especially in supporters' statements, the notion of limiting science politically was generally and clearly rejected. In this respect, framing the issue as an 'ethical' or 'moral' issue, or, more precisely, as an issue of '*individual* ethical/moral beliefs', was decisive and needs some elaboration.

Almost unanimously, the MPs found the quality of debate to be of a high quality. The decision to hold a *free vote* was of symbolic value for this opinion. "The House has shown itself at its best today. We have had an informed debate in the absence of party politics. It has been entirely about moral, ethical and medical issues" (Michael Fabricant, HC, Dec 15, 2000, col. 919). Many MPs on either side, while stressing their personal convictions, emphasised the fact that their party members were free to vote as they deemed fit, that they "fully accept the validity of arguments advanced by other hon. Members with different points of view" (Edward Leigh, HC, Dec 19, 2000, col. 258), and that they were "pleased that all sides of arguments are being put responsibly and in a considered manner" (Howard Stoate, ibid., col. 231). Also some stressed that they were following their own personal views and made "absolutely no claim to speak for anybody else" (David Wilshire, ibid., col. 246). It can be concluded that for many in the Houses the 'ethical' aspect of the new research meant that each and everyone (in the Houses and elsewhere) has their own perspective, speaks for themselves, and respects the others' views, that is, also those of disagreeing MPs. Thus, MPs inherently refused to make (definite) judgements when confronted with an issue framed as 'ethical'. An MP voting in favour of the new research brought this refusal to the point, quoting from a letter she wrote to a community of nuns:

"One of the things that I envy in people like the members of your community, who have a faith, is that you have the comfort of having certainty about these questions. I have to try to work out how to balance the different arguments. I am still struggling with that" (Fiona MacTaggert, HC, Dec 19, 2000, col. 247).

Hence, framing the new research purposes as 'ethical' issues meant, on the one hand, that they were subject to *individual* convictions and, on the other, it was not possible to aspire to, let alone attain, certainty or consensus of judgement and opinion. Indeed, any *certainty* about judgements appeared as 'moral imposition' or 'fundamentalism' in the eyes of supporters: "I find it very difficult when hon. Member's personal deeply felt opinions—which I respect—form the framework for denying other peoples opportunities [...] I have some difficulty when those beliefs are imposed on others" (Peter Brand (HC, Nov 17, 2000, col. 1195). Or as the conservative MP Robert Key said, "If we are not careful we become fundamentalists hanging on to a few dogmas that we do not intend to examine and which we will not give up" (ibid., col. 1216).

One effect of framing the new research as a matter of individual moral convictions was that state intervention was rejected in favour of individual decision-making: the issue was portrayed as a matter of individual moral judgement rather than a state decision. As supporters of the new research stated, it is subject to personal moral decisions and consent, a question of a right to choose whether or not to donate eggs or embryos, and whether or not to use therapies derived from embryo research. In the House of Lords' debate the following argument was put forward:

"There are three particular groups of people looking to us today to return to the right verdict on the regulations laid before us. The first group comprises many of the infertile couples who are being helped by the current legislation and *would like to be permitted by the new regulations* to donate their surplus fertilised eggs to be used to help sick people" (Baroness Walmsley, HL, Jan 22, 2001, col. 101, emphasis added).

Also the issue of whether or not to use 'therapies' derived from the new research was framed as an individual decision as a safeguard against 'moral imposition'. Similar to egg or embryo donation, accepting and applying 'treatment' remains subject to 'consent', upholding the 'right' to choose:

"Accepting treatment is a personal decision. It is quite reasonable that some people already do not accept certain treatments [...]. The whole point about the regulations is that they are permissive [...]. They will not force people to accept the treatment" (Dr Evan Harris, HC, Dec 19, 2000, col. 253).

This alleged 'right' to choose in cases of donating or using 'therapies', as both these quotes suggest, likewise implies that it is the state's duty to provide such a choice. Within this frame, any general restriction of scientific endeavour must appear as a 'moral imposition' or as 'fundamentalist', and, hence, as denying the option of a choice. As Yvette Cooper made clear: "If parliament votes against these regulations, hon. Members will deny couples the *choice to donate* their spare embryos to stem cell research for spinal injury or stroke" (Yvette Cooper, ibid., col. 214, emphasis added).

Thus, also in what is referred to as 'ethical' arguments, the imperative of scientific progress was the underlying assumption, because allowing ongoing research was presented as the pre-requisite for an 'ethical viewpoint' and an individual 'ethical decision'. 'Ethics' implied a personal decision on options provided by scientific research. And, after all, 'objective' science outweighed irreconcilable 'ethical' views for both supporters and opponents of the new research purposes: "No matter where one stands on the subject, it is quite clear that there are moral and ethical arguments on both sides of the debate. It is also clear that the scientific arguments are all on one side, which is to extend the Human Fertilisation and Embryology Act 1990", maintained Peter Brand (HC, Nov 17, 2000, col. 1193), while Edward Leigh, an opponent of the new regulation, stated:

> "I fully accept the validity of arguments advanced by other hon. Members with different points of view, but I point out for hon. Members who are concerned about the moral and ethical dilemmas that much scientific opinion suggests that adult stem cell research may be superior to embryo research in future" (HC, Dec 19, 2000, col. 258)

Summary: Politics or Ethics?

Against the background of embryo research that was already permitted for certain research purposes, opponents of 'therapeutic' cloning had a hard time to make their case in British parliamentary debates. Indeed, framing the issue as a problem of the 'status of the embryo' was the safety net for pro-research MPs, because it helped to interpret the new research as not new. Secondly, it was important to problematise the issue in terms of 'ethics' or, more precisely, in terms of *individual moral beliefs*, and as an issue of individual choice and rights. Within the British parliamentary debates, supporters of the new research disqualified general limits to embryonic stem cell research as a 'moral imposition' and as illegitimate intervention in a right to choose. Potential donors to the research or 'patients' profiting from it were constructed as consumers who decide along the lines of their personal moral convictions. While 'ethics' played an integral part in the parliamentary debates, parliament reclaimed the classical feature of state neutrality towards values, beliefs, and opinions and refrained from state-juridical limitations of research. General limits on research were disqualified as 'moral impositions' on the grounds of individual autonomy and freedom regarding personal moral believes. 'Ethics' was an issue of *individual* decisions and beliefs that was *distinct* from pure science

On the other hand, science was presented as inevitably 'progressing' in the world, either in Britain or elsewhere, and was therefore considered as an issue beyond problematisations or regulation and control. Despite disagreements in the British Parliament on whether or not to allow 'therapeutic' cloning, we find a striking agreement on the authoritative force of science and on the imperative of scientific progress. The fact that scientific 'progress' is set on par with societal or individual 'progress'—and that both are moral goods—were indisputable basic assumptions likewise for problematisation of the 'ethical' status of the embryo. Different opinions concerning the new research purposes were framed as being based on individual 'ethical' perspectives, whereas scientific 'progress', although highly uncertain and speculative, appeared as an irrefutable fact or, indeed, as a prerequisite for 'ethical' decisions. Problematisation of 'ethical issues' therefore could not counter but rather underpinned the call for ongoing scientific research.

The debate in both Houses was concluded with votes in favour of the new research purposes, leading to a relaxation of regulation governing embryo research. The House of Commons was divided with 366 ayes and 174 noes on 19th December 2000. On 22nd January 2001 the House of Lords approved the changes. The Human Fertilisation and Embryology (Research Purposes) Regulations (HFE Research Purposes Act 2001) came into force on 31st January 2001, introducing three additional purposes for which the HFEA may grant licences for research involving embryos according to its rules of procedure.

3.9 Conclusion: 'Great British Science'

The policy debates on reprogenetics, more precisely on research cloning, in Britain began immediately after the birth of the sheep Dolly had been announced in February 1997. High ranking policy bodies, such as the STC, the HFEA, or the HGAC were involved in the debates from the start. The British government was very proactive as regards the initiation and proliferation of the debate, renewedly postponing a decision and demanding more debate. Official bodies as well as the government focused on what was called 'public concerns' over scientific developments and the potentially restricting effects they could have on scientific 'progress', which could endanger the economic well-being of the whole nation. 'Unease' within the public was presented as deriving from ill-informedness as well as the overemphasis of risks within public debates, leading to a decline in public trust in science. Apart from a traditional deficit model of the public, a 'deficit of trust' model was at work, as Irwin (2006) called it. Lack of public trust and information were problematised as a potential threat to progress in science. Hence, to resolve this problem, the STC promoted public education to bring about public participation in an 'informed and more rational debate' guided by experts.

Against the background of this problematisation it was not surprising that, following the STC report, a public consultation on 'cloning issues' took place. This consultation was again guided by the twofold deficit model (a deficit of knowledge and a deficit of trust) and was directed at the 'improvement of informed public debate'. As regards the implicit problematisation of public debate, the Cloning Working Group consultation of the HFEA and the HGAC suggested a certain mode of discussion of and a specific attitude to 'embryo research': questioning of 'embryo research' was dismissed because it was 'a thing of the past', asserting that the issue was indisputable, and prescribing a general acceptance of embryo research in order to qualify as an 'informed' participant in the debate.

Both these consultations were based on the *imperative of scientific progress*, that is, on the assumption that science inevitably advances and will continue to do so in the future as well as assuming that scientific development was generally desirable. Despite an obvious awareness of the uncertainty of scientific results, the 'possible potential' of scientific endeavour (i.e. the so-called 'benefits') as well as science itself were beyond problematisation. Instead, and in such a framing quite reasonably so, they were to be communicated *to* and not debated *with* the public. Especially the fact that the report of the CWG public consultation was only affected to limited extent by the actual consultation obviously underpins this interpretation.

Apart from the acknowledgement of a need to discuss and regulate scientific research, an important feature of the British policy debate was the presentation of science and scientific development as inevitable and even as the prerequisite for the well-being of the nation as well as—as has been shown in the analysis of the parliamentary debates—a precondition for individual 'ethical' decisions and choices. Scientific development, however, was beyond the realm of choice. The imperative of scientific progress was at work, which the creator of Dolly the sheep, Ian Wilmut, had stressed already in 1997: "If we hadn't made the breakthrough somebody else would; the technology is out there" (as quoted in The Guardian, Feb 26, 1997).

Science and expertise formed the (not so) hidden agenda, which was presented as neutral and objective and principally beyond problematisation and politicisation. Also the integration of 'ethical' issues or public concerns did not imply that the assumed neutrality of scientific endeavour be questioned, but in fact tended to strengthen the objective-emotional divide, which several authors of ST&S studies had criticised (cf. e.g. Carr/Levidow 1997; Wynne 2001). What was presented as the scientific (as opposed to the 'ethical') dimension of the matter was not reflected on as a result of integrating 'ethics'. The strict separation of the two fields actually thwarted any reflection on scientific endeavour.

The statement by the Chief Medical Officer's Expert Advisory Group introduced a new problem: the problem of 'serious diseases' in Great Britain. By this move, 'therapeutic' cloning was transformed into a 'solution', and what is more, into a solution without alternatives. 'Problems' of cloning were transformed into second-order problems, which were divided again into 'technical problems' and 'ethical concerns about the status of the embryo'. While the former were referred to the expertise of science, the latter were referred to the area of 'irreconcilable personal moral conviction'. Such problematisation of the new research purposes, that is, as an issue involving 'serious diseases', on the one hand, and as an issue of 'personal morals', on the other, prevailed likewise in the parliamentary debates. Supporters of the changes to the law and their framing of 'therapeutic' cloning as an 'ethical' issue of 'personal moral convictions' gained dominance in the debates, localizing disagreements or conflicting views in the domain of 'individual morals' and thus outside the realm of political contestation or, indeed, intervention. Particularly the framing of the issues in terms of 'ethics' helped *not* to problematise or reflect upon science more generally. While 'ethics' was an integral part in the parliamentary debates, parliament reclaimed the classical feature of state neutrality towards values, beliefs, and opinions and refrained from state-juridical limitations of research for reasons of moral neutrality. General boundaries for research were disqualified as 'moral impositions' because they were considered an infringement on individual autonomy and freedom regarding personal moral beliefs. Indeed, supporters presented state intervention and, in general, the restriction of scientific research as a 'moral imposition' contradicting the freedom of the individual. The parliamentary decision therefore secured that science could proceed and, at the same time, signalled that public concerns were integrated: even if only by establishing a basis for personal moral decisions to be made on whether or not to donate embryos or whether or not to undergo therapies in accordance with one's own personal ethics and beliefs. The decision to legalise 'therapeutic' cloning was therefore made so that 'research would not be hindered', signalling, at the same time, that public concerns were being taken seriously.

Nonetheless, as was already mentioned, the often referred to *concerns* of the public were not voiced in a forum themselves and therefore seem fabricated, and the 'debates' referred to in the reports had a fictitious character. A newspaper analysis reveals that opposition or concerns towards or the 'overemphasis of risks' of the new research were highly exaggerated. A reason for this might have been the *anticipation* of concerns or critique after the GMO conflicts or the BSE debacle. Both these controversies had seriously challenged trust in science policy-making and in science itself. As Irwin (2006) stated, they additionally led to a shift in focus in science policy towards public trust. In this context, to restore public trust through openness, transparency, integration, and public participation became a key element in science policy. Therefore the fact that the government, from the very beginning, problematised 'public concerns' and 'lack of public trust' in the policy debates on 'therapeutic' cloning, in which conflict or opposition was rather low, can be interpreted as a precautionary measure to forestall conflict. Especially the organization of a public consultation explicitly directed at public 'mistrust' supports this interpretation.

As regards 'therapeutic' cloning we can hence conclude that the government took up a pro-active stance *stimulating* an (institutionalized) discourse on the pros and cons of the new research, to prevent conflicts before they arose. As the analysis of the different processes has shown, these discourses were not directed at the resolution of an existing conflict but at the structuring of a debate: their goal was to set the "framework for engagement" (Irwin 2006, 316) and to pre-set problematisations against the background of the imperative of scientific progress that was beyond question. This becomes evident also in the fact that the government constantly rejected the recommendations made in the reports (except for the one of the CMO), postponed a decision in favour of even more debate on the issue, and established more expert advisory bodies—underpinned by a repeated assertion of a need to reassure public trust.

This can only be partly explained by a more general tendency towards more discursive and more 'inclusive' forms of public debate and participation. Nor does public involvement signal a turning-away from a 'commitment to progress through science' as the strength of the imperative of scientific progress has shown. Rather, to involve the public was an attempt to "create a more open and reflective culture where new scientific possibilities can be fully realized" (Irwin 2006, 308) and to secure 'science's licence to practice'. In this regard the individualistic understanding of 'ethics' was decisive: It was led by the attempt to generate and promote the informed individual who responsibly chooses between options—'neutrally' provided by inevitably progressing science—against the background of her or his personal ethical convictions. The goal is to generate the "confident consumer" (Irwin/Michael 2003, 52) in a science-led society, who 'freely chooses' to donate the material needed by science or to use 'therapies' provided by it and who takes responsibility for her/his own choices. Attaining public trust seemed to be the most important goal and was to be met by allowing different perspectives as well as providing options to act accordingly: In effect, 'public trust' was to be reconstituted through respect for concerns and providing options, including the option to not use the research. The integration of 'ethical' questions or concerns functioned according to mechanisms of the market in which science produces a range of options from which individuals can choose according to their personal preferences. In this regard the stimulation of (public) discourse was the key. Discourse, it can be concluded, was a means of signalling that concerns, different perspectives, or 'moral attitudes' were respected, while the framework for engagement was pre-structured in terms of the imperative of scientific progress.

4 Germany: The Debate on Embryo Protection

In Germany, debates about the deregulation of the Embryo Protection Act (*Embryonenschutzgesetz*, EschG) leading to the reduction of embryo protection appeared around the mid-1990s, when medico-scientific actors asked for permission to perform a new techno-scientific practice, i.e. pre-implantation genetic diagnosis (PGD). The debate gained new momentum in 2000 when scientists announced that they planned to import embryonic stem cells for research purposes. As both activities were not explicitly mentioned in the act, disputes arose on whether or not German law prohibited PGD as well as whether or not the import of embryonic stem cells would be legally possible. The EschG does not mention PGD, as this practice did not exist at the time the law was drafted. The conflict, however, arose from the fact that the law mandates that the egg cells fertilized in-vitro (a precondition to perform PGD) have to be transferred to the woman from whom the eggs originated, as well as the fact that the EschG defines an embryo as 'any fertilized egg as well as any totipotent cell'. Proponents of PGD argued that the cells to be tested would not be totipotent—and thus not an embryo according to the law—which would allow the test. They also argued that PGD would help to avoid abortions, as it would guarantee that only those embryos with a negative test result would be implanted and thus the abortion of 'diseased' foetuses could be circumvented. Opponents, on the other hand, said that PGD was directed at the selection and destruction of those embryos with positive test results—a scenario which the act prohibits. As regards permission to research on imported embryonic stem cells, the legal situation was rather complicated: The EschG absolutely prohibits the destruction of an embryo and, therefore, the derivation of embryonic stem cells. However, the announcement to *import* embryonic stem cells by some scientists apparently laid bare a legal void: It was argued that, in the definition of an embryo, the act implicitly includes 'any totipotent cell' and excludes pluripotent embryonic stem cells. Others argued that imports would clearly violate at least the spirit of the EschG. A more detailed analysis of the conflicts and debates in Germany on PGD and embryonic stem cell research will, in the following, be preceded by brief account of the historical, legal, and discursive context in which they arose.

4.1 Historical Background

The following chapters will briefly introduce the German Constitution as well as Abortion Law, focusing particularly on the legal and discursive impacts they had on the debates or conflicts under study. The German Constitution is the basic legal framework for the organisation of the state as well as for the drafting, implementation, and interpretation of law. As such, it played a role in the debates and conflict on the EschG. References to constitutional articles, most prominently the basic rights on human dignity, the right to life, and the right to freedom of research, frequently appeared in the German debates. The fact that Germany has a written Constitution, which is the normative framework of the state, makes it differ significantly with Great Britain, which does not have a central constitutional document. Instead, the British Constitution consists of unwritten and written sources, which are technically not different to ordinary law.[65] References to German abortion law also played an important role as discursive devices to legitimate especially PGD and also embryonic stem cell research. Lastly, the debates and conflicts that preceded the implementation of the Embryo Protection Act in 1990 will be discussed with special emphasis on framing and problematisation processes, and the contents of the act itself explained. This brief introduction will be followed by a discussion of public conflicts surrounding 'bioethics' and biomedicine, which were relevant for the development of debates on PGD and embryonic stem cell research and that led to a formation of a rather strong movement against new biomedical practices as well as against bioethics in Germany. Compared to the situation in Great Britain, especially the existence of visible and broad protest against and opposition to 'bioethics' and biomedical practices is an important difference.

The German Constitution

The German *Grundgesetz* (GG)[66] was enacted in 1949 against the background of World War II and especially Nazi crimes. It was particularly

65 Cf. McEldoney, John. *Memorandum presented to the House of Lords Select Committee on the British Constitution*, at: http://www.parliament.the-stationery-office.co.uk/pa/ld200203/ldselect/ldconst/168/16809.htm; Aug 10, 2007.

66 Basic Law for the Federal Republic of Germany (*Grundgesetz*), translation provided by Intern. Nations, at: http://www.iuscomp.org/gla/statutes/GG.htm, Apr 2, 2006.

meant to form a bulwark against the repetition of Nazi terror and excessive state-power and is distinctive for post-war anti-totalitarian and democratic German identity (Frankenberg 2005, 5). Although provisionally enacted in 1949 until Germany was reunified and could draft a new Constitution, it was adopted as the Constitution for the "whole German people" including the federal states of the former German Democratic Republic in October 1990 (Frankenberg 2005, 4). The Constitutional Court is the supreme interpreter of German law. It "shall rule on the interpretation of the Basic Law" (Art. 92(1)) in the case of conflicting views on scope or interpretation of constitutional rights as well as monitor the implementation of new law with regard to its compliance with the Constitution.

The Basic Rights (Arts. 1 to 19) are subjective rights that take precedence over any laws and norms and are binding for state powers (Frankenberg 2005, 6). This means that nobody, including the legislature, has these constitutional guarantees at their disposal, and that restrictions are subject to immense limitations: the basic rights are subject to a legal proviso (Art. 19(1)), according to which their 'essence may in no case be affected' (Art. 19(2)) (cf. Frankenberg 2005, 6). However, Article 1 on "the inviolability of human dignity" and on "human rights" is explicitly excluded from amendments (Art. 79(3))[67], as especially this first and supreme normative principle was meant to prevent any repetition of totalitarianism (Frankenberg 2005, 5). Thus, Art. 1 is absolute and eternal, meaning that it can never be amended or balanced against other constitutional rights or ordinary law. Furthermore, "human dignity" is not a constitutional *right* among others but an essential *quality* of human beings, which must be protected by the state as a constitutional guarantee. In contrast, Art. 2(2) on the "right to life" may be restricted through law—as in cases of self-defence—while Art. 5(3) on the "right to freedom of research" may not. In the conflicts in Germany over PGD and embryonic stem cell research as well as in debates on bioethics policy more generally, the constitutional rights played important roles as legal and discursive resources. Especially the Basic Rights: Art. 1. on the "protection of human dignity" and on "human rights"; Art. 2(2) on the "right to life and bodily integrity"; Art. 5(1) on the "freedom of expression"; and Art.5(3) on the "right to freedom of art,

67 Articles on the division of the Federation, the basic institutional principles (democracy, federal organisation), or the protection of the constitutional order (Art. 20) are also excluded.

science, research, and teaching" were often referred to in the debates and became the object of conflict-ridden argument.

The German Constitution also regulates the federal and democratic organisation of the state, likewise including the legislative competences of the *Bund* or the *Länder*. At least one change of the Constitution was important for policy-making regarding biomedical issues. In 1994, legislative competence with regard to artificial fertilisation, the examination of inheritable 'information', as well as regulation of tissue and organ transplantation (Art. 74(26)) was transferred from the *Länder* to the *Bund*. Matters such as fertility treatment or genetic testing were, from then on, issues of "concurrent legislative power of the federation" (Art. 72). This meant that the *Länder* had the power to legislate insofar as the *Bund* has not exercised its legislative power by enacting a law. The *Bund*, however, can legislate on these matters in order to establish equal living conditions throughout the territory (Art. 72(2)). Prior to this, as was the case in drafting and implementing the Embryo Protection Act, federal law could only overrule *Länder* law in matters of health policy—and thus create legal unity as regards fertility treatment or embryo protection—through penal law.

German Abortion Law: An 'Unborn Life' and a 'Conflict'

References to the German Abortion Law frequently appeared in the policy debates concerning PGD and embryonic stem cell research. Liberal regulation of abortion was often identified with soft embryo protection, which served as discursive tool to legitimate PGD or embryonic stem cell research. In this respect, especially framing abortion as an issue concerning the embryo as well as framing it in terms of the 'conflict of the pregnant woman' was important. [68]

Abortion in Germany is regulated through in the Penal Code (*Strafgesetzbuch*, StGB, Art. 218, Art. 219).[69] Defining abortion as a punishable felony, the act was, practically since its implementation at the close of the 19th century, an issue of public conflict, entailing calls for liberalisation as well as calls, especially by women's groups, for the elimination of the

68 To be sure, I do not presuppose an essential relationship between abortion and PGD or ES cell research, but rather seek to highlight the discursive interlacing of the issues in so far as abortion practice and law formed important reference points in the debates on reprogenetic research.

69 StGB: *Strafgesetzbuch*, at: http://bundesrecht.juris.de/stgb/index.html, Dec 13, 2006.

clause (Marx Ferree et al. 2002, 26ff). The act underwent several reforms, especially under the German Nazi regime, which introduced a distinction between so-called "worthy" and "unworthy lives", forbidding abortion of the former and demanding abortion of the latter. In 1949 the pre-war Art. 218 was re-enacted. It remained more or less uncontested until the 1960s and 1970s when it was re-politicised (ibid., 28). Several suggestions concerning the regulation of abortion—including the so-called *Fristenlösung*[70] and the so-called *Indikationslösung*[71]—were introduced, until in 1975 the West German Constitutional Court ruled that abortion law must take into account that the 'unborn life' needed protection (Jasanoff 2005, 159). In 1976 the West German Parliament passed a law ruling that women can get an abortion if a doctor certified indications, including criminal, medical, social, and embryo-pathological indications (Marx Ferree et al. 2002, 35).[72] East Germany, on the other hand, had adopted permissive abortion law, granting women abortion on demand during the first trimester in 1972 (ibid., 33).

In the German reunification process in the beginning of the 1990s, abortion law became one of the most difficult issues to be settled (Jasanoff 2005, 159).[73] The conflicts especially over these issues were too strong to be settled in the unification treaty, which therefore postponed a decision on abortion until after a new *Bundestag* had been elected. The treaty specified that "by legally securing the claims of women, particularly for counselling and social support", the new law "better guarantees the protection of the unborn life" (as quoted in Marx Ferree et al. 2002, 41). Unwanted pregnancies were framed as "conflicts that needed management" (ibid.) in the treaty, leaving specification of who was to manage the conflict undefined. Although it emphasised that "the claims of women" (ibid.) should be secured—including the demands of East German women to implement a

70 The *Fristenlösung* means that abortion should be allowed in the first trimester.

71 *Indikationslösung* means that abortion is allowed for certain indications.

72 However, most abortions at the time relied on social indications. Still, the application of this practice differed immensely between the *Länder* and especially physicians in Bavaria faced criminal penalties for failure to comply with the law. In the so-called Memmingen Trials a physician who performed abortions was sentenced to a prison term (cf. Jasanoff 2005, 160).

73 As Jasanoff writes, Christian Democrats managed to frame East German abortion law "as just another of the exploitative, dehumanizing practices of a failed and discredited regime" (2005, 161). Abortion was functionalized in the attempt to build a coherent German identity that completely rejected communism (ibid.).

liberal abortion law— the framing of unwanted pregnancies in terms of the 'protection of unborn life' and 'conflict-management' localized the debate on a new law within the West German constitutional framework (ibid.). In 1993, the Constitutional Court overturned an appeal for a new, more liberal law, on the grounds that it did not sufficiently protect 'the unborn life'. It insisted on abortion remaining a criminal act unless termination of the pregnancy was justified in cases of rape, incest, or threat to the life of the mother (ibid., 42). The court ruled that counselling should obligatorily precede abortion and be directed at preserving the 'life of the unborn' although open in the way in which it was to be carried out. Women who have undergone counselling were to be exempt from prosecution. In 1995, a *Fristenlösung*, allowing abortion in the first trimester, augmented by obligatory counselling and a clause on indications, was implemented (ibid., 42f). At the same time, embryo-pathological indications were abolished, whereas criminal and medical indications remained unchanged (Jasanoff 2005, 161).

The court's ruling provided the frame for the drafting and implementation of today's abortion law, which affirms the moral and legal *subject* status of the 'unborn life' as well as corroborating the framing of unwanted pregnancies as 'conflict situations' (cf. Duden 1993; 2002, 237). The notion of 'pregnancy conflict' and its proclaimed need for 'management' prepared the prescription of pregnancy-conflict counselling in cases where women sought an abortion without any of the legal indications. While German abortion practice today appears as if it were abortion on demand, professional counselling directed at the 'protection of the unborn life' is the precondition for the fulfilment of this demand.[74] As Graumann (2002a, 111) writes, feminists have indeed rejected the discussion about the 'protection of the unborn life', which they understood to be a spurious conflict that obscured the 'real' issue: Women were locked in the role of motherhood that disregarded their basic rights. The women's movement interpreted the new reproductive technologies, including prenatal or pre-implantation genetic diagnosis of embryos, that transformed pregnancy into a medical and technical issue, as further strategies for disempowering women (ibid., 112).

Today, Art. 218 of the penal law rules that abortion is illegal but not punishable under certain circumstances: If the woman underwent counsel-

---

74 Dominique Memmi called the state's abstention from explicit prohibition in favour of delegating decision-making to the counselled, individual women "delegated biopolitics" (*la bio-politique déléguée*) (Memmi 2005; cf. also Herrmann/Könninger 2008).

ling at least three days before the abortion took place; if a doctor carried out the abortion; and if nidation took place no more than 12 weeks beforehand. The counselling process is generally regulated in Art. 219 StGB and rendered more precise in the *Schwangerschaftskonfliktgesetz* (Pregnancy Conflict Act; SchKG 1992). Art. 219 mandates that the counselling process "serves to protect unborn life" and must be directed at "encouraging the woman to continue the pregnancy" (author's translation). The second clause of the abortion law (Art. 218a (2)) regulates *legal* indications for abortion: If abortion was carried out in order to prevent "a threat to the life or severe impairment of the physical or psychological well-being of the pregnant woman" (author's translation) and, provided that it was done with the woman's consent, abortion is legal in Germany with no temporal restrictions (Art. 218a (2)). As 'embryo-pathological indications' have been completely removed from the act, abortion is illegal for reasons concerning the embryo. However, while embryo-pathological indications for abortion are today technically prohibited, *de facto* they are regularly performed, being subsumed under the 'threat to the life and health of women' clause. Today, nobody seriously questions that women should have access to abortion, despite the existence of pro-life views and actors in Germany.

On the grounds of a discourse on 'unborn life', abortion policy and the discourses surrounding it were important reference points in the following debates on policy-making on PGD or embryonic stem cell research. Under certain circumstances, the fact that abortion is prohibited but not punishable in German law was often held against the argument that embryo research would violate the constitutionally granted right to life of the embryo. A common argument of pro embryo research actors was that the embryo in-vitro would be better protected than that in vivo. The same argument was often held against the view that PGD would violate the EschG: pro-PGD actors argued that the technology was, in principle, comparable to abortion, except that the 'activity' was performed earlier (e.g. GfH 1995). Also abortion law served as model for suggestions to regulate PGD on a case-by-case basis incorporating counselling processes, as supporters of the practice suggested. Especially feminist actors against PGD, on the other hand, argued that, according to the law, abortion is only allowed in a situation of pregnancy *conflict* of the woman—which was not the case if the embryo is in a petri dish and in a situation where the woman clearly wants to get pregnant. Therefore, PGD and abortion were incomparable, according to these actors (e.g. Mildenberger 2002). Indeed, what was strongly opposed to in the 1990s' abortion debates—i.e. the framing of abortion in terms of a pregnancy *conflict*—now became an important reference point for feminist opposition to PGD.

4.2 The German Embryo Protection Act: History and Content

History

The 1990 German Embryo Protection Act is the result of a year-long seesaw between different medical or research institutions, the government and parliament, as well as the *Bund* and the *Länder* (cf. also Richardt 2003). After the birth the first IVF baby in 1978, different commissions and institutions at a federal and state level had dealt with the issues of in-vitro fertilisation, genetic technology, and embryo research. While genetic technology was then not (yet) a public issue, at the time Germany faced the growing significance of science and technology and especially nuclear power as well as an increased scepticism towards the same (cf. Aretz 1999; Radkau 1988). The nuclear power conflicts of the 1970s and 1980s, which triggered mass demonstrations and, to an extent, militant protest, had shown that highly developed technologies bear great potential for the politicisation thereof and therefore brought about a situation that policy-makers wanted to avoid for genetics (Radkau 1988, 343f). The first expert hearings organised by the ministry of research and technology dealing with the 'ethical implications' of IVF and genetic technology were held in 1979 and 1983. They explicitly sought to counteract problematisation and politicisation of genetic technologies in addition to structuring and framing public debate on the issue before it actually took place (Herrmann 2006). Hearings and consultations with experts were organised in order to prevent the "public attitude from becoming emotionalised towards science" (author's translation, Herwig/BMFT 1980, XII) and to avert "fundamental political debates" (author's translation, BMFT 1984, 98) such as those of the nuclear power conflicts. The research minister asserted that it was therefore vital to prove to the public that science and science policy-makers 'act responsibly' to "preserve a measure of freedom for science far-removed from any intervention from outside", including that of the state (author's translation, ibid., 98). The first two hearings, however, were not directed at legislative recommendations but at evaluating "the necessity or the disadvantages of state intervention in the form of guidelines or laws" (author's translation, Herwig/BMFT 1980, XI). It was concluded that it was unnecessary to provide legal regulations or guidelines (BMFT 1984, 97). Altogether, the first couple of years after the birth of the first IVF baby were marked by the framing of new biotechnological developments as expert issues, which did not give reason to problematisation and politicisation. What was problematised, however, were anticipated public debates that could lead to 'fundamental politicisation' and thus endanger the development of science. Therefore, an "ethical-philosophical debate" led by experts should be accompanied by a "public acceptance debate" through which public debate was to be rationalised before it actually took place (ibid., 98f).

In the mid-1980s, however, pressure on policy-makers to demonstrate their intention to regulate the new practices grew stronger (Gill 1991b, 108) and two commissions with the task of preparing legislative processes were established: the ministerial Benda Commission (BMFT 1985) and the parliamentary Enquête Commission on "Chances and Risks of Genetic Technology" (EK DB et al. 1987) (*Enquête-Kommission 'Chancen und Risiken der Gentechnik', EK Gentechnik*). The Benda Commission[75] had been established in 1984 by the ministries of justice and of research and technology in order to evaluate the "ethical and legal aspects of the application of new reproductive procedures" (BMFT 1985) and to make recommendations for conduct. The Benda Commission dealt mainly with artificial reproductive technologies (ARTs), whereas the *EK Gentechnik* (set up in 1984) focused on genetic technologies, including human and agricultural genetics. This separation had been criticised, for example, by feminist actors as well as the Greens (cf. articles in: Die Grünen im Bundestag/AK Frauenpolitik e.V. 1986). The two technologies, so it was argued, should be analysed together, and the fact they were separated was interpreted as a strategy to circumvent critique of the one (IVF) infringing on the other (genetics) (Gill 1991b, 183). With regard to implementing safety guidelines on genetics, the *EK Gentechnik* was described as a "lesson in political non-decision" (author's translation, Gill 1991a; 1991b; cf. also BT DR 10/1581; Palfner 2003, 29ff). [76]

---

75 Named after its chair, the former Constitutional Court President Ernst Benda.

76 Not only did it take several years until safety guidelines on genetic technology were introduced, but it was also criticised that the Act on Genetic Technology (*Gentechnikgesetz*, GenTG), which came into force in 1990 (and was further deregulated in 1995), did

The first years in which new bodies or processes were introduced to deal with new research were marked by the aim of conflict prevention (Herrmann 2006). The main goals concerning the function and effects of the various bodies and processes were to protect professional self-regulation and prevent restrictive state intervention and societal conflicts via the rationalisation of an anticipated public debate. The actors, very explicitly, wanted to set the goals, terms, topics, and the disciplinary authority of a prospective debate in order to avoid public conflicts similar to those experienced in the 1970s and 1980s over nuclear power plants (Interview D1, Oct 20, 2005). The topics were generally framed as expert issues, which, however, had to be 'explained' to the public before conflicts arose. In subsequent public discourses on genetics and, more generally, on science and technology issues, problematisations in terms of 'chances and risks' gained dominance. By framing them in this way, the development of the technology itself—and certainly questions concerning the desirability of the so-called 'chances' of the technology—as well as its social or political implications could not be problematised (cf. separate opinion to the report by the Green EK-members: in EK DB et al 1987, 314ff). The integration of 'ethics' or 'ethical' issues, on the other hand, was a device to prove the responsibility of the scientific community in order to prevent conflicts, or, as was the case in the Benda Commission, to find legal limits to research and application (Interview D6, Dec 1, 2005; cf. Herrmann 2006; Braun et al. 2008).

The Benda Commission on IVF, genetic testing, and embryo research published its report in 1985 (BMFT 1985). The report focused on constitutional/legal problematisations of new technological options and on jurisdictional recommendations that were deemed necessary. The problems of IVF, genetic testing, and embryo research were also framed in terms of chances and risks, that is, in terms of their 'great potential' (ibid.), such as the fulfilment of a wish to have a child, on the one hand, and the risks and dangers of the new technology, that is, the fact that IVF embryos exist outside the mother and therefore being open to manipulation and embryo research, on the other. A second frame effective in the Benda report was that of a conflict between the 'freedom of research and the basic rights of a third person'. While the research under consideration was understood to be principally worth funding because of the 'chances' it promised, its limits

not seriously impede the release of genetically modified plants and that the public was excluded (Dolata 1996).

would be established through the basic rights of others, the report concluded. Hence, the 'freedom of research' would have to be limited where constitutionally secured basic values were affected. These would include protecting the autonomy of the people concerned, protecting the well-being of the child, or 'protecting the status of the embryo as it deserves' (ibid., 5).

The recommendations made in the Benda report mainly referred to questions of family policy and embryo protection. The questions included such problems as who is allowed to use reproductive technologies, what will the status of the father be, or what is the status of a child born after IVF. The Benda Commission recommended that embryo research be allowed to a limited extent. While the Benda report made rather clear recommendations, they were not simply translated into law. On the contrary, what followed were years of conflict between the *Bund* and the *Länder*, medico-scientific professionals and research institutions, as well as public actors over the question of whether or not regulate ARTs and embryo research and, if so, how (cf. Richardt 2003).

Some of the conflicts over the regulation of reproductive technologies resulted from formal legal requirements concerning the area of health policy. As health policy—which included ARTs—was the responsibility of the federal states, national law could only be introduced as criminal law, which would overrule also state law, and, as such, had to be passed in the *Bundestag* as well as the *Bundesrat*. Whereas the federal government (a coalition between CDU/CSU and FDP) and especially the federal ministry of research (responsible for the drafting of an act) were in favour of the new technologies and especially embryo research, the Social Democrat and Green Party opposition was critical of them. At the *Länder* level, conservatives and liberals favoured restrictive policy and social democrats and greens were opposed to embryo research. It was unpredictable whether or not the *Bundesrat* would affirm a decision made by the *Bundestag* (ibid.).

Also within the public sphere there was opposition to embryo research as well as artificial reproductive technologies more generally. Especially feminist opposition was significant (Bradish et al. 1989; Die Grünen im Bundestag/AK Frauenpolitik e.V. 1986). At the societal level, in the 1980s IVF was a much more controversial and more expansive issue than was represented in the Benda Commission. Especially the feminist movement was very concerned about the social and political consequences of a technology that, in their view, was a misogynist device to 'totally control' reproduction and transfer it to the sphere of profit and market interests (cf. articles in: Die Grünen im Bundestag/AK Frauenpolitik e.V. 1986). Feminists argued against a technological fix and, instead, demanded a more ecological, social, and respectful interaction with the environment and human beings. They discussed many different aspects: among others, the meaning of the desire to have a child, of unwanted childlessness, questions of identity, family policy, violence against women, or the contrast between population policy in Western countries as opposed to anti-natalist population control in countries of the South. Moreover, feminists not only opposed separating genetic and reproductive technologies, which became evident in the establishment of the two separate advisory commissions, but also separating human and agricultural genetics. In brief, feminist critique of genetic and reproductive technologies included a general socio-political critique of the technologies themselves, of societal and political conditions of technological development and their consequences, as well as of the structures of science policy-making. The congress "Women Against Genetic and Reproductive Technologies" (author's translation, ibid.) demanded a halt on public financing of these technologies and was explicitly opposed to "non-public ethics commissions" (i.e. Benda Commission), which only "sanctified" technologies (Resolution 1986, 18). 'Ethics', especially what was identified as 'expert ethics', was suspected of being a "legitimation ritual" (author's translation, Prätorius 1985, 92) because of its potential to narrow down the scope of debate and to block more general socio-political discussions.[77] 'Ethics' was interpreted as a hegemonic political and strategic tool to confine the debate on IVF, limit human genetics to technical and legal details, and exclude more general socio-political voices and debates.

After the Benda Commission had finished its work, several guidelines, drafts, and white papers were published, revised, and newly published by the federal as well as the state governments and by medical or research organisations (several of them are published in Hülsmann/Koch 1990). In response to the Benda report, the conservative-liberal coalition government produced a first white paper that generally prohibited embryo research, allowing, however, exceptions subject to approval by professional authorities. This rather liberal approach was criticised, among other reasons, because it was doubted that the medico-scientific profession could

77 This suspicion was also related to the growth of professional bioethics, which was interpreted as being 'legitimation ethics' for the natural sciences.

successfully control such research (cf. Richardt 2003, 88). Whereas research and medical organisations claimed that they were able to self-regulate IVF or embryo research and sought to prevent state regulation, the federal and state governments were locked in a conflict over competences in legislative power. The federal government issued a draft white paper that allowed limited embryo research, whereas the state of Bavaria introduced a draft of a Reproductive Medicine Act (*Fortpflanzungsmedizingesetz*, FmedG) that entailed the prohibition of embryo research and the transfer of regulatory responsibilities from the medical profession to the state. As a response, the medical profession withdrew its claim to be allowed to carry out embryo research in order to avoid the loss of regulatory competences, thereby demonstrating its ability to regulate the matter in an ethically responsible way (ibid., 104).

In 1990, the German government published the last draft of an Embryo Protection Act representing a compromise between older versions and a FmedG, following a proposal put forward by the state of Bavaria. Embryo research should be prohibited and no governmental agencies were introduced to oversee ARTs. ARTs remained the responsibility of the medical profession. Along party lines, both houses agreed to the draft of the EschG, which came into force in January 1991 (EschG 1990). To the present day, Germany does not have an act regulating reproductive medicine, but its practice is subject to the self-regulation of the medical profession (cf. BÄK 2006).

The Act

The German Embryo Protection Act is criminal law and prohibits the "improper use of reproduction technology" (EschG 1990). As the name of the act suggests, its key role is *embryo protection*. It provides a legal framework for practising artificial fertilisation and handling embryos created via such technologies (cf. Kollek 2000, 189). The goal of the act, as Kollek asserted, is to secure the treatment of unintended childlessness and, at the same time, prevent the creation of 'surplus' embryos and the misuse of embryos (ibid.). The act prohibits egg cell 'donation' as well as the attempt to fertilize an egg cell for purposes other than to establish a pregnancy. It mandates that not more than three eggs may be fertilized within one treatment cycle and that all fertilized eggs have to be transferred. It also prohibits surrogate motherhood. It declares that the use of the human embryo for purposes not serving its preservation, that is, purposes other than the establishment of pregnancy, as well as gender selection are punishable by law. Furthermore, it rules out that an egg cell can be fertilized without the consent of those from whom egg and sperm had been taken. The woman receiving the fertilized egg must consent to the transfer. The act also mandates that only physicians may carry out artificial reproductive techniques (EschG 1990).

The EschG forbids that human germ cells be artificially altered in any way as well as cloning techniques be utilised, such as creating a human embryo with identical 'genetic information' to that of another embryo, a fetus, a human being, a dead person, or hybrids or chimaerae. It defines an embryo as "the human egg cell, fertilised and capable of developing from the time of fusion of the nuclei, and, further, each totipotent cell removed from an embryo that is assumed to be able to divide and to develop into an individual under the appropriate conditions" (ESchG 1990, Section 8(1)).

As long as technical options of performing interventions in embryos and usages of the latter were limited, it was a straightforward matter that they were protected from being used in ways other than to establish pregnancy (Kollek 2000, 189). However, the EschG became the target of critique because practices such as pre-implantation genetic diagnosis (PGD) were introduced and the use of embryos for medical or research applications grew to be scientifically attractive (ibid.). Several aspects must be highlighted concerning conflicts on reprogenetic issues that arose in the mid-1990s, that is, PGD and embryonic stem cell research: First of all, the EschG does not regulate or even mention PGD because this practice did not exist at the time of its drafting and implementation. It was developed at the end of the 1980s/beginning of the 1990s. It comprises the genetic testing of an embryo created via in-vitro fertilisation for so-called 'genetic or chromosomal defects', whereby only those with negative test results are transferred to the woman and the others discarded (ibid., 27ff). Thus, the practice inevitably entails the production of 'surplus embryos', that is, of embryos not used to establish a pregnancy—a scenario prohibited by the EschG. However, as the cells used for a PGD are scientifically defined as 'pluripotent' and not 'totipotent', they do not, according to some, fall under the act.

Furthermore, even though the act actually prohibits the derivation of embryonic stem cells (which means the destruction of the embryo) it does not explicitly prohibit their import and use. This was considered a legal void, which was crucial at the end of the 1990s, when several German researchers announced that they wanted to import embryonic stem cells for research. However, although German law is restricted to German territory (StGB, Art. 3), Art. 7 Abs. 2 of the German penal code (*Strafgesetzbuch*, StGB) states that an activity proving to be an offence in the country in which it is performed is punishable also in Germany. Additionally, some argued that German imports of ES cells must not be related to their derivation. In this view, if at all, imports had to be restricted to those ES cells that were produced legally in the country from where they were imported and independent of the German demand for them.

Before I go into the details of the conflicts surrounding PGD, embryo research, and imports of embryonic stem cells, which commenced in the mid-1990s, I will first discuss problematisation of 'bioethics-discourse' connected to the formation of an anti-bioethics alliance in Germany at the end of the 1980s in the following section.

4.3 Formation of an Anti-Bioethics Alliance: Problematising Bioethics Discourses.

The years between the end of the 1980s and the mid-1990s were marked by public conflicts on 'bioethics' and biomedicine policy in Germany preceding problematisation of the Embryo Protection Act. The conflicts were prompted by inviting the Australian bioethicist Peter Singer to Germany in 1989 and by the unauthorized publication of the "European Convention on Human Rights and Biomedicine"—or Bioethics Convention for short— in 1994 (cf. Braun 2000a; Braun/Herrmann 2000). These events contributed to the formation of a rather broad but diverse civil society protest alliance against biomedical practices and bioethics discourse. The existence of an alliance opposed to (utilitarian) bioethics discourses and biomedical developments in Germany marks a major difference to the situation in the UK, where opposition or techno-scepticism was rather fragmented, on a greatly reduced scale, and less organised.

Against the Erosion of Human Rights and Human Dignity

Following the invitation of the Australian bioethicist Peter Singer to speak in Germany in 1989 by a disability organisation, a debate arose on whether discourse on the 'right to life of newborn (disabled) babies' should be tolerated or not. Even though Singer, who openly supported infanticide in the case of so-called 'severely damaged newborns' and 'euthanasia' (cf. Singer 1982), faced strong opposition in the German public, he also found support. Especially groups and individuals of the disability movement or disability organisations campaigned, sometimes violently, against his public appearances. The main points of critique in Singer's views were that he applied utilitarian ethics to (disabled) human beings and he rejected their right to life. Singer was regarded as continuing Nazi propaganda by openly denying disabled newborns the right to life and by promoting 'euthanasia' (cf. Jordan 1999). Singer's opponents argued that a fundamental right to life was denied if the discourse on this issue continued and therefore demanded that it be stopped (Tolmein 1990a).[78] Singer's supporters, for their part, argued for freedom of speech and called for discussing his views openly (cf. Tolmein 1990b). Numerous German university professors declared their solidarity with Singer in the "Declaration of German Philosophers on the so-called 'Singer Affair'" (Erklärung [1989] 1991). They also demanded an open discussion on Singer's proposals even in cases of disagreement with his views (cf. Tolmein 1993). Nonetheless, the anti-Singer campaigners succeeded in stopping most of Singer's presentations in Germany at the time, and a discussion of Singer's theses remained a non-option (Interview D5, Noc 30, 2005). With reference to Peter Singer and to utilitarian bioethics or debates on 'euthanasia', a conflict resulted that was partly due to different views on the function and effects of discourse: Singer's opponents framed the problem as a fundamental attack on constitutional rights, such as the right to life, by starting a discourse on such issues. A fundamental claim of the anti-bioethics alliance was therefore *no discourse on Singer's theses* as a form of opposition. Other actors, however, problematised the oppositional arguments, also in terms of the constitution, as attacks on the right to freedom of speech irrespective of content (Erklärung [1989] 1991).

Opposition to bioethics discourse in Germany escalated and altered its appearance in the mid-1990s within the context of a citizens' initiative publishing an unauthorised translation of a draft of a European Bioethics

78 Cf. also flyer of the AStA Braunschweig, at http://www.forum-bioethik.de/BioethikN1_3_6.html, May 18, 2007.

Convention in May 1994 (Frankfurter Rundschau, May 25, 1994).[79] In 1991, the Parliamentary Assembly of the Council of Europe commissioned the European *Comité Directeur pour la Bioéthique* (CDBI) to draft a 'European Bioethics Convention' to protect the 'human being' in the context of biomedical research (on the history of the Bioethics Convention cf. Braun 2000b).[80] Because of strong controversies, especially on the issues of embryo research and research on so-called incapacitated persons, it took several years and revisions before the CDBI decided, in November 1996, on a version that the ministerial committee of the Council of Europe agreed to (cf. Emmrich 1999). In the meantime, the Bioethics Convention was renamed the *European Convention on Human Rights and Biomedicine*[81]) and was submitted for signature and ratification in April 1997 by the member states (Emmrich 1999). At present (Dec 2008), 34 member states have signed the convention with 21 of them having also ratified it.[82] At the time, the work of the CDBI, the drafting process, and members of the commission were little or not at all known to the public, and neither were the reasons for drafting this kind of convention nor the intention behind it (Braun 1998; 2000b). The German citizen initiative responsible for the publication in 1994 achieved this by way of "democratic indiscretion" (author's translation, Fuchs 1999, 270). They started a campaign to make the contents known to the public and to stop Germany from signing or even ratifying it. The initiative was supported by numerous organisations and individuals (Fuchs 1998a; several statements are printed in Wunder/Neuer-Miebach 1998a). What followed was a controversy on the contents of the convention and, in particular, on the secretiveness of the whole drafting process (cf. Emmrich 1997).[83]

---

79 Internationale Initiative gegen die geplante Bioethik-Konvention. [Citizens against the proposed Bioethics Convention].

80 Braun (2000b) provides a critical reflection of the Bioethics Convention and, more generally, bioethics as regards human rights and human dignity.

81 The full name of the Conventions is: "Convention for the Protection of Human Rights and Dignity of the Human Being with regard to the Application of Biology and Medicine: Convention on Human Rights and Biomedicine", at: http://conventions.coe.int/Treaty/en/Treaties/Html/164.htm, Dec 20, 2008.

82 A list of signatories to the Convention can be found at: http://conventions.coe.int/Treaty/Commun/ChercheSig.asp?NT=164&CM=&DF=&CL=ENG; Dec 20, 2008.

83 The debate is documented in Paul (1998). On the reactions to the publication cf. http://www.uni-wuerzburg.de/gbpaed/mixed/misc/bioethik/reaktion.txt; Apr 12, 2005.

The main point of critique was that utilitarian ethics were applied to human beings in the draft, permitting research on people who were unable to give their consent and including research that would not benefit them personally (Braun 2000b, 205). Activists also criticised that, in general, embryo research was permitted up to the 14th day of the embryo's development and that the issue producing embryos for research purposes was only vaguely defined (cf. ibid., 238ff). They recognised the danger of the draft convention annulling the German Embryo Protection Act via the back door. The 'Convention on Human Rights and Biomedicine', contrary to its name, would not secure but in fact endanger human rights, (Höfling 1998; cf. also Braun 2000b). The *Arbeitskreis zur Erforschung der Euthanasie-Geschichte* (Study group for research on the history of euthanasia), an initiative inciting public reflection of Nazi eugenics and euthanasia and working against a renewal of Nazi medicine, published the so-called Grafeneck Declaration in October 1995, in which the "dangers of today's bioethics" manifested in the convention were discussed (author's translation, Grafenecker Erklärung [1995] 1998).[84] As the declaration relates, bioethics and bioethical balancing of costs and benefits, especially concerning research on people who were unable to give their consent, disagrees with and relativises human as well as basic rights. In the declaration the following statement quite accurately summarises the critique of the alliance against the Bioethics Convention:

> "With great concern we observe that bioethics is not conceived as an instrument to preserve human rights but, on the contrary, abandons the foundation of human rights, disobeys historical experience, and relinquishes human rights protection of the individual to the expedient balancing of values" (author's translation, Grafenecker Erklärung [1995] 1998, 183).

An important antithetical point of reference for German protest against bioethics was German Nazi history, and the protest alliance warned not only against its recurrence but also against sustaining a rationality that was directed at "enhancing deficient human nature" (author's translation, Wunder/Neuer-Miebach 1998b, 11). Bioethical references to "super-ordinate interests" or to economical considerations of "the health of human kind" reminded the alliance of the rhetoric and practices of Nazi doctors and scientists (author's translation, Grafenecker Erklärung [1995] 1998,

84 The Grafenecker Erklärung can also be found at: http://www.fuente.de/bioethik/grfneck1.htm; Jan 24, 2005.

191f). The Bioethics Convention, as Michael Emmrich (1999, 342) wrote, meant a total renunciation of the Nuremberg Code and of the lessons Germany should have learned.

Human rights and human dignity were important points of reference for the alliance, and most statements by critics of the convention included paragraphs on the ways it represented an attack on and annulled human rights and dignity. It was argued that the convention was based on the divisibility and gradable application of human rights—for the benefit of science—instead of a universal idea of indivisible human rights and inviolable and presuppositionless human dignity (Braun 2000b). Instead, the convention introduced a system in which particular human beings could be classified as *incapacitated* and made available for research purposes. The fact that Art. 2 of the Bioethics Convention codified the principle of balancing interests between the individual and science or society was seen as clearly violating the principle of human rights (ibid., 250). Altogether, despite differing views in other areas, the groups and individuals critical of the convention agreed on the assumption that it would annul or at least seriously endanger human rights and dignity rather than secure them (cf. Emmrich 1999). This critique was also extended into parliament, where an inter-factional group of MPs was formed against the ratification of the convention in Germany under the heading *Bündnis für Menschenwürde* (Alliance for Human Dignity) (Interview D5, Nov 30, 2005; cf. Braun 2000b, 210).

In sum, the anti-bioethics alliance understood that biomedical development as well as bioethics discourse contradicted basic human rights and human dignity. The problem was framed as a *conflict between basic rights and (utilitarian) bioethics*, which, in their view, meant a denial of human rights and dignity and entailed the danger of repeating Nazi crimes. The critique of the protest alliance was highly successful: because of the strong opposition within the German public and in parliament, Germany has still not signed the convention today (March 2009). While the critique has since abated, at least as regards publicity, several groups, citizen's initiatives, and organisations fostering it still exist today and are present especially in the internet.[85]

---

85 Such as the Interessengemeinschaft Kritische Bioethik Deutschland at: http://www.kritischebioethik.de/deutschland_home.html; May 20, 2007; a website critical of the bioethics convention at: http://www.bioethik-konvention.de; April 1, 2005; the Forum Bioethik at; http://www.forum-bioethik.de; May 20, 2007, or BioSkop publishing a

A Demand for Public Discourse

However, the protest was not only directed at substantial questions, but also strongly problematised the intransparency, expert bias, and clandestine manner of the drafting process (Fuchs 1998b). It additionally objected to the fact that the contents of the conventions were declared to be confidential (Fuchs 1999, 270). The protest alliance feared that German MPs could do the same in the ratification process. Hence, an essential claim of the German anti-bioethics alliance was the demand for a broad public debate on the issues of the convention, which, in their view, concerned each and everyone. *Was alle angeht müssen alle entscheiden*[86] was the slogan of the protest alliance demanding participation already in the process of will formation, and it likewise insisted on transparency in the debates on ratification.[87] As the protest alliance interpreted it, the problem—at least to an extent—stemmed from the lack of public discourse, and public discourse was presented as (part of) the solution. However, the alliance did not demand discourse in general but a certain kind of discourse: they did not simply demand more public discourse, but called for more discourse highlighting sceptical viewpoints and potentially strengthening opposition, a discourse that was not restricted to experts and that was grounded in human dignity and human rights (Interview D5, Nov 30, 2005). It is apparent that the demand for more discourse implied that critics were neither ready to discuss the issue with advocates of Singer-type bioethics nor to debate over the issue of certain individuals having the right to life and others not. On the contrary, the demand for more discourse was inseparably bound to an attempt to re-frame the discourse in terms of the fundamental right to life and the principle of human dignity. Utilitarian bioethics discourses were still the object of fierce critique.

The attempt to re-frame the discourse on biomedical issues was an essential feature of bioethics-critical discourses. A result was that the *Institut Mensch, Ethik und Wissenschaft* (IMEW, Institute of Humankind, Ethics, and Science) came into operation in 2002 (founded in 1998).[88] The goals of the

---

critical quarterly journal on the developments in the life sciences at: http://www.bioskop-forum.de; May 20, 2007.

86 What concerns everybody must be decided on by everyone.

87 Cf. Bürger gegen Bioethik, at: http://www.fuente.de/bioethik; Feb 8, 2007.

88 Cf. Institut Mensch Ethik und Wissenschaften: Decree on the establishment of the IMEW at: http://www.imew.de/imew.php/cat/78/title/Gruendungserklaerung, Jan 10, 2007.

IMEW are to conduct independent research in medical ethics and to establish a dialogue between science and society "founded on respect".[89] Likewise, its aim is to establish a counter-discourse to bioethics.

Another result of the protests against bioethics and biomedical research and practices was to establish a parliamentary Enquête Commission on Law and Ethics of Modern Medicine (*Enquête-Kommission Recht und Ethik der modernen Medizin*; EK REM) in 2000 (cf. 4.7). For activists as well as several MPs, such a commission seemed to be an appropriate way to create awareness in the public on the issues of the convention and guarantee public participation, as well as draw attention to public and extensive parliamentary debates (Interview D5, Nov 30, 2005). In particular, the alliance believed that it would allow techno-sceptical views to enter the arena of decision-making and strengthen opposition to the convention, as the spokesman of a citizens' initiative declared (Südwest Presse Ulm, Feb 3, 2000). The protest against bioethics now problematised a non-democratic decision-making processes in exclusive expert discourses behind closed doors, supporting an attack on basic human rights—a problematisation that (at least partly) the protest groups believed could be resolved by means of more inclusive, more open discourses. Very importantly, however, the demand for more discourse was not in itself understood to be a solution, but was linked to a demand for the restriction and control of scientific research and application: They clearly insisted on "strict controls" and "slowing down" for biomedical research, which was to be grounded in human rights (author's translation, Grafenecker Erklärung [1995] 1998), and "a general prohibition of research on disabled people. [...] After all, this is what we demand" (author's translation, Südwest Presse Ulm, Feb 3, 2000).

Altogether, the members of the anti-bioethics alliance in Germany were manifold, including disability groups, feminists, pro-life groups, citizen's initiatives, and MPs, that is, groups that might not have agreed on all issues, such as abortion[90], but were united in their opposition to utilitarian bioethics policy and discourse as well as to particular biomedical practices, such as embryo research or PGD. The existence of a large and diverse

89 Cf. Institut Mensch Ethik und Wissenschaften: About Us. http://www.imew.de/imew.php/ cat/74/title/About_the_Institute, Jan 10, 2007.

90 In rare unity, pro-life and feminist groups joined together in mutual disagreement over new biomedical practices with an unspoken agreement to not touch upon the question of abortion. In an interview, a German pro-life MP said that "it is not the case that we have dropped this topic [abortion, slh] altogether, but we have decided to debate the issue in a different setting" (author's translation, Interview D5, Nov 30, 2005).

anti-bioethics alliance in the 1990s in Germany marks a notable difference to the situation in the UK, where groups or individual actors disapproving of new biomedical developments such as embryonic stem cell research were much rarer and less unified and often pro-life,[91] or at least identified as such. In summary, the German anti-bioethics alliance criticised substantial and procedural aspects of policymaking in biomedical developments, suggesting that more public discourse would help strengthen opposition and finally lead to restrictive policy and strict controls. Such an idea was alien to the UK.

4.4 The Debate on Embryo Protection: Setting the Agenda

Against the background of a rather strong anti-bioethics/biomedicine alliance that wanted to foreclose unrestricted ongoing biomedical developments and stop bioethics discourses, another debate began in Germany in the mid-1990s that was initiated by individuals and organisations representing professional research interests and not by protest groups. This new debate dealt with, more specifically, the Embryo Protection Act and, in particular, new reprogenetic practices, such as pre-implantation genetics diagnosis (PGD), embryonic stem cell research, and imports of embryonic stem cells.

The German Embryo Protection Act (EschG) was relatively uncontested until the mid-1990s, when a discussion on drafting and implementing an act on reproductive medicine started anew within the *Länder* (cf. Beier et al. 1994). However, the *Länder* initiative became redundant in 1994, when the German *Bundestag* and *Bundesrat* amended the Basic Law and referred regulatory competences on health issues, including reproductive medicine, from the *Länder* to the *Bund*. Therefore, whereas the legal regulation of reproductive technologies was previously subject to *Länder* law and to the self-regulatory competences of the professional medical chambers of the *Länder*,[92] after the amendment of the basic law, the implementation of a *federal* law on reproductive medicine became likely, which could possibly have replaced or restricted self-regulatory competences of the medical profession in favour of state control. At the same time, discus-

91 Personal discussion with members of the British NGO *The CornerHouse*.

92 The *Landesärtzekammern* (Chamber of Doctors of the *Länder*) are bodies of public law and inhibit indirect state authority.

sions on a new law also provided the possibility for softening embryo protection. In this situation two professional organisations, the German Research Association (*Deutsche Forschungsgemeinschaft*, DFG) and the German Society for Human Genetics (*Deutsche Gesellschaft für Humangenetik*, GfH) issued statements in which they problematised the prohibition of PGD and embryo research. Accompanying the first official application to be allowed to perform a PGD, they problematised 'over-regulation' of research through legal restrictions (DFG 1996a, 1010) and a conflict between 'legal restriction and the interest of future parents' (GfH 1995). As a result, both promoted more professional autonomy as a solution.

The Freedom of Research and Its 'Irrational Impediment' through Regulation

In a position paper of 1995, the German Society for Human Genetics (GfH)[93] asserted that the EschG prohibited PGD on *totipotent* cells as this would equal the destruction of an embryo. A similar test on so-called *pluripotent* embryonic cells was said to be generally allowed if it served the establishment of a pregnancy (GfH 1995, 420). Still, as PGD requires in-vitro fertilisation (IVF) of an embryo—which according to professional guidelines is only allowed for *infertile, married* couples—it is expressly forbidden for solely diagnostic purposes (ibid.). These legal and professional restrictions would, however, contradict the interests of future parents, such as avoiding a 'tentative pregnancy', abortion, and the wish to use PGD to clarify genetic risks for severe 'defects' in a child' (ibid.). PGD should generally be open to all couples who wish to clarify such a 'genetic risk'. According to the GfH, it was essential in this respect to establish comprehensive professional guidelines that restrict the practice to 'medical problems' (ibid.).

A second attempt to problematise the prohibition of PGD, embryo research, and restrictive regulation of scientific research more generally came from the German Research Foundation (DFG), one of Germany's leading associations of scientists and a prominent public funding agency for research, in May 1996. In its plea for the freedom of research, the DFG (1996a)[94] presented its view on "(over)regulation" of research through law,

93 GfH webpage can be found at: http://www.gfhev.de/index.htm; Jan 1, 2007.

94 The article (DFG 1996a) used here for the quotation is a shortened version of the "Plädoyer" that includes only the chapters of the book dealing with embryo research and general issues. The book version also includes chapters on other science fields, such

in particular in the EschG, and on the presumed negative effects for "freedom of research" (ibid., 1008f). The DFG complained that, in drafting the EschG, the legislator did not take into account that "adequate rules and guidelines in scientific communities or professional organisations" (author's translation, ibid. 1010) already existed in order to circumvent the 'misuse' of IVF and embryo research. In comparison to the professional guidelines, the EschG would 'block' embryo research in the field of reproductive medicine in Germany altogether (ibid. 1019f). Without explicitly demanding or promoting embryo research, the DFG stated—on the possible violation of basic rights—that "stepping down on the issue of the basic right to life in view of the basic right to freedom of research could ultimately be necessary to save or preserve life by developing therapies" (author's translation, ibid., 1015). Among other reasons, this quote prompted the Association of German Scientists (*Vereinigung Deutscher Wissenschaftler*, VDW) to publish a reply, in which it argued that the DFG's statement would pave the way to allow research on people without obtaining their consent and without benefiting them personally (VDW 1996, 4).[95] The DFG spoke of "legislative hypocrisy" (author's translation, DFG 1996a., 1020) in the prohibition of embryo research whilst the use of the results of the same research performed in other countries was legal. Similarly, the DFG discussed the prohibition of PGD: In contrast to the UK, allowing this research under certain circumstances and controlled by the HFEA, German law "disadvantaged" patients who would (have to) go abroad to benefit from new medical practices, that is, if they could afford to (ibid., 1023). Additionally, the DFG believed that German researchers might leave the country and increasingly work abroad. This would have serious effects for research in Germany and could likewise present a legal problem for German researchers. 'Over-regulation' was problematised as the result of misinterpreting science on the part of the legislator due to the under-representation of scientific perspectives in the law-making process. In fact, the DFG claimed, "the self-restriction of the legislator" in the area of science and research in favour of "self-control of science" was crucial for securing "freedom of research" in Germany (author's translation, ibid., 1011). Consequentially, self-control of science and—if necessary—the

as animal experimentation, genetic technology, nature and monument protection, or data protection (DFG 1996b).

95 Cf. also VDW webpage: http://www.vdw-ev.de; May 30, 2007.

inclusion of science and expertise in the law-making process were presented as adequate solutions.

Thus, a debate on PGD and embryo research began with problematisation of legislative activity concerning scientific matters in general and embryo protection in particular, triggered by professional organisations such as the GfH and the DFG. Framed as being caused by a deficit of scientific knowledge within law-making processes, legal restrictions for scientific endeavour were problematised as endangering Germany as a base for science, limiting patient rights and the rights of science. Taking a deficit model as a basis by assuming that the legislator did not adequately reflect scientific knowledge in the legislative process, scientists and professional organisations framed a problem that ensued, in principle, from a conflict between the *freedom of research and its unjustified legal restriction.* Additionally, legal restrictions of embryo research or PGD were framed as contravening *the needs and wishes of patients* and thus, at a more general level, as conflicting with *the public interest.* In this regard, the professional interest in performing PGD was identified with the public interest.

A Case for PGD: Legal Restrictions versus Ethical Obligations

While the two papers introduced above remained at the level of statements, Germany saw its first concrete case for carrying out a PGD in 1995: Two professors from Lübeck medical school asked their clinical ethics committee for permission to perform a PGD in one special case. Referring to both statements discussed above, the ethics committee turned down the application because the EschG stipulates that embryos can only be created to establish a pregnancy and must be transferred to the woman's womb. This would, "*expressis verbis*", prohibit PGD (Oehmichen 1999a, 110; original emphasis) as it anticipates the destruction of positively tested embryos. But while the committee turned down the application for *legal* reasons, at the same time it doubted whether the EschG was "still ethically justified" in this respect (author's translation, ibid.). "Ethically", the committee stated, "the performance of a PID [sic!] was certainly warranted for this couple" (ibid., 107).[96] In this instance, the issue of PGD was framed primarily as a 'health issue' in that it helped avoid a troublesome abortion in case of the possible transfer of 'genetic defects' to the offspring (ibid.

96 English abstract prefixed to the statement.

109). The woman involved in this special case, as the committee emphasised, "asked of her own accord" (author's translation, ibid., 108) if PGD was available in Lübeck. The committee concluded that, provided that the parents 'autonomously' decide for a PGD after having received detailed genetic counselling and comprehensive instructions about the technology, its risks, and possible necessary repetitions, the decision had to be accepted from an 'ethical' point of view (ibid., 111).

To sum up, the ethics committee problematised a conflict of *legal restrictions* versus *the ethical obligation to help couples and accepting the autonomous decisions of the same.* This problematisation was underpinned by the implicit problematisation of an apparent 'contradiction' between German abortion practice and the EschG, where it was suggested that an abortion, after a positive prenatal test, would be "possible" or even "necessary" (ibid. 109) and would provide greater burdens for the woman/couple than a PGD—a 'contradiction' interpreted as being 'ethically doubtful'. The committee obviously argued along the lines of assuming that abortion or destruction of a 'genetically burdened' embryo would *necessarily* follow after a positive test result. However, it must be noted that German abortion law, contrary to the statements, does not allow abortion for 'embryo-pathological' reasons, despite the fact that 'medical' abortions after prenatal testing are regularly performed. Nevertheless, the 'contradiction' argument underpinned problematisation of legal restrictions in terms of a 'public/ individual—and not merely professional—interest' by permitting PGD, and on these grounds PGD was framed as an 'ethical obligation' towards individual couples.

Altogether, all three statements discussed above presented a problem, which was depicted either as a conflict of the 'constitutional right to freedom of research' versus 'inadequate legal restrictions on it' or as a conflict of 'ethics' versus 'law'. It was stated that legal restrictions of PGD were 'ethically doubtful' or 'wrong' and that 'ethical obligations' should overrule legal provisions. Legal restrictions were framed as contradicting the public/individual interest and thus as being 'unethical'. Despite an emphasis on the public/individual interest, professional expertise was to remain the guidepost for decisions on PGD: Genetic counselling and expert instruction should precede the 'autonomous' decision of the concerned couple. In effect, the ethics committee suggested that, similar to German abortion law, expert-guided, individual decision-making should replace legal restrictions.

'Totipotency' versus 'Pluripotency'

All three statements discussed above triggered a discussion in Germany on the question of whether or not the EschG prohibited PGD. In this period, the debate was dominated by a scientific framing of PGD as an issue concerning 'totipotency versus pluripotency' and therefore as an issue concerning the *scientific status* of the biopsied cell. Similar to the invention of the 'pre-embryo' in the 1980s in the UK (cf. Spallone 1989) or the differentiation between 'therapeutic' and reproductive cloning in the British stem cell debates (cf. also Herrmann 2003), the distinction between 'pluripotent' and 'totipotent' embryonic cells was a device to classify one activity—that is, to test 'pluripotent' cells—as legitimate by delegitimating the other—that is, to test 'totipotent' cells. According to scientific criteria, a pluripotent embryonic cell is neither assumed to be capable of developing into a human being nor an embryo according to the EschG and therefore can be examined and destroyed. Although the scientific problem of when 'totipotency' of human embryonic cells ends and 'pluripotency' starts is rather old, as Kollek (2000, 65ff) writes, no answer has been found as yet—even in scientific terms. At the time of drafting and implementing the EschG at the end of the 1980s, a clear definition or temporal limitation of 'totipotency' did not exist (ibid., 65). The notion was integrated into the act without defining within the law what it exactly meant (ibid., 65), and the differentiation between the two 'classes' of embryonic cells can only be made by scientists on a theoretical basis. Experiments proving the 'non-totipotency' of so-called pluripotent cells were certainly illegal in Germany because they would be, primarily, experiments on the woman: A test directed at determining whether or not a cell is able to develop into a human being would necessarily include that the cell be transferred to a woman, a scenario the act clearly prohibits. Also, experiments would fall under the cloning paragraph within the EschG (1990, Section 6, (1)).

While the notions of 'totipotency' and 'pluripotency' were rather vague and controversial, the ensuing debate on whether or not PGD was legal and/or acceptable was framed in terms of an opposition between the two. The most influential publications suggesting the legality of PGD were a report by the Bioethics Committee of the federal state *Rheinland-Pfalz* (the Caesar Commission) (Caesar 1999)[97] and a publication of the draft guidelines for performing PGD by the German Medical Association (*Bundes-*

97 Cf. http://cms.justiz.rlp.de/systemstatic/justiz/pdf/Bioethik.pdf, Oct 30, 2007.

ärztekammer, BÄK) (BÄK 2001).[98] Both publications argued that PGD was not forbidden by the EschG as long as the biopsied cell was not totipotent, and in this way avoided a more general questioning of PGD practice: Indeed, PGD practice was "understood to be an innovation in reproductive medicine" (author's translation, Caesar 1999, thesis III, 1). Problems that might arise, so the commission stated, "differ somewhat, depending on whether the diagnosis is performed on totipotent or non-totipotent cells" (author's translation, ibid.).

Public Reservations and 'Progress' of Science

Apart from a general acceptance of the practice of PGD (on pluripotent cells), the Caesar Commission problematised that "rapid developments" in reproductive medicine face "reservations in Germany" (author's translation, ibid., preface). This problematisation was based on the assumption that it "is imperative to find ways to use progress in biomedicine" while "fear of PGD must be rationally assessed with regard to its validity" (author's translation, ibid.). 'Fears' were interpreted as problematic and needed to be addressed and 'rationalised' in order to secure progress in science. Similar to the British example, the Caesar Commission problematised the 'validity' and the possible effects of 'fears' (or as in the British case, 'mistrust') in the new practice rather than problematise the practice itself. On the grounds of a perceived value contradiction compared to abortion law, it framed PGD as a conflict between protecting the 'rights of the embryo' and protecting the interests of the future mother/parents by not necessitating her/them to have a 'diseased' child (ibid., preface). The commission recommended that the best way to balance these values was to introduce professional standards for performing PGD instead of state regulation or legal restriction: Expert guidance was to replace state rule.

"In the case of pre-implantation genetic diagnosis, state protection of the embryo does not have to be put into effect by means of penal law. In fact, it can also be achieved by standardising the requirements for performing PGD, by which means

98 The Bundesärztekammer (BÄK) is the central body of medical professionals in the German system of self-administration. It represents its members in matters relating to professional policy. It plays an active role in the opinion-forming process concerning federal health policy and legal procedures. One of its main tasks is to regulate the ethical and professional obligations of doctors to each other and to their patients and to regulate professional medical duties and principles for medical safety.

one can much more effectively take future parents' interests into account, which are likewise relevant in the balancing process" (author's translation, ibid., thesis II, 7).

About half a year later, the BÄK published a discussion paper on guidelines for the use of PGD (BÄK 2001). The 'ethical discussion' on PGD, so the BÄK stated, essentially encompassed the problem that an embryo created in a petri dish might not be transferred into the uterus if the of a test result is positive (ibid., 157). Similar to the Caesar report, the BÄK's statement was based on the assumption that PGD was generally a practice used to avoid abortion and was demanded by concerned couples. Therefore, it was the responsibility of the future parents to decide (ibid., 158). While "this serious, fundamental ethical decision […] is primarily the responsibility of the concerned couple" (author's translation, ibid., 158), the BÄK considered that the lack of 'legal security' for doctors necessarily involved in PGD was the main problem. The BÄK suggested in its concluding remarks: "If the majority of society is in favour of pre-implantation genetic diagnosis", the best way to achieve "legal security" and guarantee "high standards of protection" (author's translation, ibid., 158) was to implement professional guidelines for case-by-case examination. Guidelines, however, should arbitrate between two virtues: they should not "inappropriately restrict the possibilities of modern diagnostics" and, at the same time, take "the need to protect human life" seriously and respect the interests of those people, who, in "great fear of having a genetically-determined seriously ill child, are in danger of damaging their own health" (author's translation, ibid., 162).

Hence, the BÄK not only adopted the framing of PGD that was introduced by the Caesar Commission but also went beyond it. PGD was framed as a conflict between the need to protect the embryo, on the one hand, and the interests of the concerned couple/mother, on the other. However, the BÄK went further by interpreting the 'interests of the concerned couple' as being linked to their personal health and thereby implicitly countered arguments that frame PGD as a practice of selecting embryos according to 'quality standards'. Likewise, the BÄK implicitly referred to abortion law, which allows abortion in the case of health risks for the mother and prohibits abortion for embryo-pathological reasons (cf. 4.1). Therefore, framing the issue in this way is again based on the assumption that PGD is a practice that makes it possible to avoid an 'inevitable' abortion, presupposing that 'genetically diseased' embryos would be aborted in any case.

Problematisation of PGD in terms of an ethical conflict of *the right of the embryo versus future parents' personal health interests* had particular consequences for suggestions on how to govern the practice: Firstly, it underpinned the idea that the 'fundamental ethical decision' was the future parents' responsibility and not that of the state or the medical profession. Secondly, however, references to *health* interests referred PGD to the domain of the medical profession, which had the duty of providing medical guidance and implementing medical standards. Similar to the Caesar Commission, which favoured professional standardization of the practice and case-by-case decision-making on its application by means of regulation through law, the BÄK (2001) suggested that professional guidelines were the best means of providing 'legal security' and 'high standards'. Individual cases should be presented to a BÄK ethics committee responsible for approvals and rejections. Professional organisations introduced a two-track solution for the governance of PGD practice: At one level, problematisation of a conflict between the embryo and the future parents led to suggestions that PGD regulation be transferred to a professional decision-making process under professional guidance. At another level, however, it was suggested that counselling processes be introduced for concerned couples, who then take responsibility for the 'final decision' on whether or not to use PGD and whether or not to have the embryo transferred to the woman (BÄK 2001, 166f; Caesar 1999, 3).

Techno-Sceptical Answers: Socio-Political Consequences and State Rule

Disagreements on whether or not the EschG prohibited PGD led some (including political officers and scientific experts) to suggest that a new act on reproductive medicine should clarify the matter by, as some maintained, expressly allowing PGD (cf. Beier 1996; Neidert 1998). Other actors, however, including the then Federal Health Minister, Andrea Fischer (Bündnis 90/Die Grünen) (cf. Fischer 2001b), the Minister of Justice, Herta Däubler-Gmelin (SPD) (Bahners et al. 2001), as well as professional and other civil society organisations, such as women's organisations or associations of disabled people, insisted that PGD was and should remain prohibited (e.g. ReproKult 2002a; ReproKult 2002c). In this situation, in which it was not even clear whether the EschG allowed PGD or not, especially the BÄK's proposal of guidelines for the use of PGD was understood as an attempt to change the direction of the debate, that is, to end the discussion on *whether or not* PGD was legal and, instead, to focus on *how* to regulate its use.

In the view of techno-sceptical actors, PGD was clearly violating the EschG because the embryos with positive test results would not be transferred but discarded—a scenario the act prohibits. Also techno-sceptical actors believed that PGD would enforce discrimination against disabled people and undermine social solidarity with them. The practice of selecting embryos according to 'quality standards' would abet eugenic tendencies within society and introduce a practice for classifying human beings as 'lives not worth living' (cf. e.g. Netzwerk 2001, Netzwerk 2006). The selective goals of PGD and the fact that PGD would inevitably lead to the creation of 'surplus embryos' have been highlighted by several techno-sceptical groups including also scientists (cf. e.g. Kollek 2000; ReproKult 2002a; VDW 1996). Feminist actors added that the practice put pressure on women to avoid the birth of a disabled child and forced them to face new decisional conflicts (ReproKult 2002c). Also PGD could only be performed together with IVF and would impose an extension of indications on IVF treatment,[99] which is associated with many psychological and physical risks for women. PGD would therefore face the same difficulties that IVF is confronted with (Deutscher Ärztinnenbund 2000). Additionally, feminist techno-sceptics argued strongly against an "instrumentalization of women's needs in the interests of research" (ReproKult 2002c, 135; cf. die tageszeitung, Apr 12, 2000). Feminists opposed the line of reasoning that PGD would support and be an answer to women's needs (ReproKult 2002c). They argued that PGD would "establish social pressure to use the technology, which would force women to face new decisional conflicts" (ibid., 136; cf. Kollek 2000, 161ff). They also were against the fact that new biomedical practices, such as PGD, increasingly transfer responsibility for the 'health' of a child to the mother/parents (ReproKult 2002d) and pleaded for a politico-social framing of the issue instead: "Restricting the argument to the level of personal needs conceals the social consequences of this technology" (ReproKult 2002c, 136). The result could be some kind of "back door to eugenics" (Graumann 2000, cf. Duster 1991). Disability rights advocates claimed that the introduction of PGD would signal to

99 IVF treatment was hitherto only indicated for fertility problems of couples and not for genetic testing of the embryo.

people with disabilities that they were not welcomed by society (Tolmein 1998). Even more so, Graumann (2000, 16) added, if whether or not to have a disabled child is a 'free choice'. Altogether, groups, organisations, or individuals opposed to PGD favoured more complex problematisations of the issues than supporters did. Techno-sceptical actors framed the issues in terms of *socio-political consequences*. Their arguments focused on the socio-political dimensions of PGD, such as the discrimination of disabled people, social pressures on women to have a healthy child, or the instrumentalisation of women's needs in the interests of science and research. Techno-sceptical civil society organisations or individuals explicitly countered arguments that maintained PGD was a practice serving the interests of the public or those of the individual woman or couple.

Another important frame employed by techno-sceptical actors can be called the *slippery slope frame*. It problematised the difficulty of controlling PGD and restricting its application to a few cases, arguing that its confinement to 'high-risk' couples or to very 'serious diseases' would be ineffective already on account of it being impossible to define these criteria. Therefore, it was inevitable that the practice be extended to encompass the treatment of more diseases and in more cases (Berg 2002; ReproKult 2002c). Also feminist and disability rights activists claimed that limiting PGD to a small number of the 'most severe diseases' was impossible, which was verified by the example of prenatal diagnosis (PND) (Netzwerk 2006). Opponents of the practices believed that, once accepted, PGD would soon turn into a wide-spread screening practice and would open the gates for other even more problematic practices such as germ line intervention (Kollek 1999; Tolmein 1998). Additionally, many of these actors claimed that the legalisation of PGD would inevitably lead to the creation of 'surplus embryos' (embryos not transferred to the woman) and could thus become a gateway for an increased production of more embryos in IVF treatment for research use. Many of these actors also suspected that PGD opened the doors for the deregulation of the EschG in order to make embryos available for research (Kollek 2001; ReproKult 2002a; c).

In sum, the issue of PGD in Germany was by no means either a well-defined or an agreed upon but rather a "wicked problem" (Fischer 2000, 128). Actors in Germany differed to a great degree on what they thought the 'problem' 'really' was. The struggle surrounding PGD was therefore not a struggle over how to solve a well-defined problem, but it was, in point of actual fact, a struggle about problematisations. It was a struggle about the meaning of PGD, about how to frame and define what is problematic about the practice, and about setting the discourse on it. Suggestions of how to 'solve' the respective problem were related to various problematisations: Techno-sceptical actors promoted legal prohibition by state rule, whereas supporters of PGD suggested that its practice be allowed and that it be regulated through measures provided by the scientific community (standardisation, counselling).

Summary: Initiation and First Phase of Debate

Whereas the 1980s were marked by a strong anti-bioethics/biomedicine discourse that turned against unrestricted biomedical 'progress' and utilitarian bioethics discourse, in the mid-1990s a new discourse emerged that promoted new biomedical practices and scientific research. This new discourse on embryo protection was not initiated by protest groups but by professional organisations and individuals who wanted to perform and support new biomedical and -scientific practices, especially PDG, at a time when the new technology could be used. At the same time, the implementation of a new federal act became likely. This situation meant that either unwanted research restrictions would be abolished or that there was the risk of losing self-regulatory competences. Indeed, especially the German Research Foundation (DFG) found self-regulation to be illegitimately constrained already. At the very beginning of this pro-research and application debate, professionals problematised a conflict between the constitutional right to freedom of research and 'illegitimate' and obstructive legal restrictions. The latter was accompanied by problematisation of a conflict between restrictive law and the interests of the future parents. The EschG was attacked from different angles: On the one hand, there was a more general demand on policy-makers for 'legislative abstinence' in science matters in favour of professional self-regulation on the basis of the constitutional right to freedom of research. If necessary, science should take a lead in regulatory processes that were based on the assumption of sound science being rational and neutral. In this respect, the *Lübeck* case seemed like an event that was used to create publicity and stimulate a public debate. At another level, PGD was framed as a matter concerning the opposition between toti- and pluripotency. While this purely conceptual and theoretical distinction became a central argument in statements by techno-optimist actors, it could only (if at all) be determined by scientific experts and was subject to change along with scientific 'progress'. This framing paved the way for some to declare that PDG on pluripotent cells was allowed in the EschG anyway and intrinsically stating that there was no need for new regulation.

On the other hand, PGD was framed as being in the individual medical interests of 'genetically burdened' couples who want a healthy child, whereas legal restrictions were defined as inadequate–and possibly 'unethical'—because they restricted couples' 'autonomous decisions'. By references to 'the majority in society' who 'want PGD', to 'high-risk couples', to patients and scientists 'who have to go abroad to use or perform the practice', or to patients who 'ask for PGD on their own accord', the issue was framed as one of public and individual—and not merely scientific—interests. In this way, legal restrictions were presented as conflicting not only with scientific standards but also with public/individual needs and rights. The demand for expert counselling and professional guidelines, however, attests to the fact that also 'individual interests' were framed in terms of scientific and medical expertise. The fact that these actors referred to 'public or individual interests' in PGD cannot conceal that actually an imperative to (allow the) use (of) science formed the basis for problematisations presented by techno-optimists.

All in all, scientific or medical actors framed PGD as a medico-scientific issue concerning the scientific 'status of the cell' as an issue that was of public and not merely of scientific interest and as a means for testing the willingness of the state to safeguard the 'freedom of research' and especially the interests of the public. This task, however, was presented in such a way as to suggest that it would be best executed and controlled within the professional system of self-organisation and -government. This framing underpinned the calls of professional organisations for new *professional* strategies concerning the governance of PGD and embryo research.

Problematisation of restricting PGD was a device for stimulating a discourse and opening up a space for demanding the relocation of regulatory competences regarding embryo protection. It was an attempt to change the status quo of governing scientific research and application. While forms of state control and state regulation were problematised, another form of government—namely professional self-government—was presented as a more rational and effective alternative. The discussion paper on guidelines for PGD practice produced by the BÄK can therefore be interpreted as an attempt to document the ability of the medical profession to successfully govern the use of this practice. The presentation of professional self-government as a more appropriate form of government, however, was accompanied by suggestions for professional counselling procedures and, eventually, *individual* decision-making processes.

Techno-sceptical answers offered more diverse and quite different problematisations, which were devoted to the social and political implications, in particular, of PGD and, in general, of science and technology. In fact, contrary to Fraser's (1989a) 'suppression hypothesis', techno-sceptical actors, comprising disability groups, feminists, or other less powerful social actors, *answered* techno-optimist attempts to change the status quo and sought to introduce a different problematisation of PGD into the debate. Their success therein, however, was only limited. Techno-sceptical actors believed that the EschG prohibited PGD and rejected calls for its legalisation on the grounds of the socio-cultural consequences, which in their view would ensue if PGD was performed in Germany. Techno-sceptics problematised PGD in terms of the negative *social consequences and conditions* of the new practice. At the same time, particularly feminist actors opposed the line of argument that the new practice would support women's needs—a line of argument they declared to be an instrumentalisation of women's needs. Techno-sceptics opposed a framing of PGD or, more generally, biomedical developments in purely scientific or medical terms, and disagreed with the proposition that professional self-regulation should play a prominent role. Instead, techno-sceptics defended restrictive state policy and rejected attacks on the EschG as unacceptable professional encroachments.

4.5 The 'Years of Biopolitics'

In Germany, 2000 and 2001 were marked by notably more discourse on biopolitics in the media, the public, and at the level of institutionalised discourse: In 2000 the German Parliament established a study commission on "Law and Ethics of Modern Medicine" (*Enquête-Kommission Recht und Ethik der modernen Medizin,* EK REM), which was under pressure to prepare exigent legislative activity concerning biomedical developments. Federal Health Minister Andrea Fischer, in preparation of a new act on reproductive medicine, organised a huge symposium, which was extensively reported upon. Finally in 2001, Chancellor Gerhard Schröder established the

German National Ethics Council (*Nationaler Ethikrat*, NER) to advise the government on questions of biomedicine.[100] Throughout the whole period, newspapers organised expert debates on biomedical issues in which also many political officials took part. We can thus justifiably speak of an *explosion of discourse* in 2000 and, even more so, in 2001 (cf. Graumann 2002b, 13).[101] The national newspaper *Die Zeit* spoke of a "year of biopolitics" (Die Zeit, Dec 28, 2000).[102] In the following I will first discuss the public debates appearing in 2000 and 2001, summarising the main arguments and problematisaisons that came up in this period, and then focus on the institutionalised debates, i.e. the Fischer Symposium, the EK REM, and the NER.

Dolly, Cloning, Embryonic Stem Cells: More Conflicts

While the conflicts surrounding PGD were still going on, in 2000 the issues that gradually began to dominate were cloning, embryonic stem cell research, and embryo research. The announcement of Dolly's birth in a national newspaper in February 1997 (Frankfurter Rundschau, Feb 26, 1997) prompted an intense debate also in Germany. All larger newspapers contained articles on Dolly, and the main issue discussed in the beginning was whether or not German law would preclude the cloning of human beings. Dolly's birth was the reason for many newspapers to initiate expert debates centring on an 'ethical' reflection of (human) cloning (cf. articles in Ach et al. 1998).[103] The national newspapers *Frankfurter Rundschau* und *Die Zeit* both organised extensive expert debates on human cloning. Whereas these debates initially agreed that human cloning should be rejected (Graumann 2002b, 26), this credo lost its force when the so-called 'therapeutic potential' of cloning connected to embryonic stem cell research and the differentiation between 'therapeutic' and 'reproductive cloning' entered

100 The symposium and the two commissions will be discussed in chapters 4.7 and 4.8.

101 Graumann analysed twelve newspapers and weekly journals and found 184 articles for the years 1995 to 1999, whereas for the years 2000 and 2001 she found 2350 articles on biopolitical questions.

102 *Die Zeit* conduced considerably to the intensity of the debate; almost every issue contained articles on biopolitical issues. The online version of *Die Zeit* has a considerable archive on articles on biopolitics. Cf. Die Zeit, webpage, at: http://www.zeit.de, Dec 30, 2005.

103 Cf. also articles in Graumann (2001), Geyer (2001), and articles at: http://www.zeit.de, Aug 25, 2004.

into the debate. Embryonic stem cell research first came on the agenda in Germany in 1999. It was initially associated with PGD in the debate, but later it overlaid disputes over PGD. Human embryonic stem cells had been isolated and cultivated for the first time in 1998 (Thomson et al. 1998); a practice that is prohibited in Germany by the EschG. Prompted by the publication of the successful transfer of a nucleus into a de-nucleated egg cell of a mammal two years ago that resulted in the creation of Dolly, discussions arose on the possibility of so-called 'therapeutic' cloning and embryonic stem cell research.

The scientific option of using embryos in research and the (potential) medical applications thereof again made the EschG a target of controversy, and again discourse was stimulated by professional scientific organisations. The first incentive triggering debate on embryonic stem cell research in Germany came from the German Research Foundation (*Deutsche Forschungsgemeinschaft*, DFG). In March 1999, the DFG published a statement on the "issue of human embryonic stem cells" (author's translation, DFG 1999), in which it discusses the pros and cons of human embryonic stem cell research as well as the legal situation in Germany concerning this type of research. In light of legislation and in view of the application of research in other countries, especially the more relaxed regulation in the UK, the DFG suggested that it might be advisable to change German legislation at some time in the future. They foresaw that this type of research would find support for its so-called 'therapeutic and diagnostic potential' in particular and, likewise, because it was allowed in other countries. The DFG said it would be 'ethically' doubtful to use research results that were developed in other countries while prohibiting the research itself in Germany. Nonetheless, the need to use so-called 'totipotent' cells posed a barrier for research because this was prohibited by the EschG. The DFG accepted, however, research on pluripotent cells. Funding of research should still be subject to "the examination of arising ethical issues" (author's translation, ibid., 6).

The DFG believed that the opinion-building process on 'ethical' and embryological questions was still in its initial stages and suggested that a broad debate on the issue take place in which the DFG would participate (DFG 1999, 6). Although there was, as yet, no need to change German law, the DFG declared that this may change in the future. In order to be ready to deal with possible conflicts arising from anticipated research developments—a situation that could make legislative changes 'necessary'—

the DFG suggested that a central commission be established for evaluating, controlling, and supervising research projects from an ethical, legal, and scientific viewpoint (ibid., 6). Whereas the DFG did not see a problem for the present, a *future* problem was anticipated, and it was framed as a conflict between research potential and restrictive legislation. A solution to this problematisation could be anticipated by discourse on 'ethical questions' while, for the time being, research was able to proceed.

Germany had its first case for the new research in August 2000. Researchers from Bonn University applied for funding of imports of and research on embryonic stem cells. A few months later in December, the then British Prime Minister Tony Blair announced that Britain intended to become the world's leading nation in biotechnology and the British Parliament amended the HFE Act and allowed so-called 'therapeutic' cloning. All these events led to a broad debate on embryo research in the German press. One of the most prominent issues discussed was the question of whether or not Germany could abstain from this research field and at the same time import results (in the form of 'therapies') of the same research from other countries. The British decision to allow 'therapeutic' cloning functioned as a catalyst in the German debate, leading to a massive increase in discourses on reprogenetics and biomedicine at the end of 2000, which lasted through 2001 in the German press and public.

Re-Framing the Debate: The 'Ethics' of Economics and Human Dignity of the Embryo

Accompanying the upsurge in discourse was a serious change at the semantic level of discourse: Around this time it is significant that we can observe an obvious narrowing down of issues and problematisations entering the debates. Before 2000, the debate encompassed a wide range of issues, such as the status and problems of children selected or designed via reprogenetic techniques; the commercialisation of the female body as a resource for 'harvesting' egg cells or embryos; the increased discrimination of and decrease of solidarity towards disabled people; and the danger of slipping into more and more applications of the new technologies, health issues, or needs and wishes of concerned couples. In contrast, at the close of the Millennium, debates in newspapers tended to focus much more narrowly on embryo protection and the 'status of the embryo' (cf. also

Graumann 2002b). Additionally, sparked off by Chancellor Gerhard Schröder, economic questions now played a prominent role in the debates.

Because embryonic stem cell research was presumably an economically more promising field than PDG, two days after the British decision on 'therapeutic' cloning the then German Chancellor Gerhard Schröder (SPD) engaged in the debate for the first time, pushing for an economically focused perspective on biomedical developments. Schröder declared that it was time to do away with "ideological blinkers" and "general prohibitions" and not to lose sight of the international market (author's translation, Die Woche, Dec 20, 2000). He said that it was now important for Germany to not lag behind international developments in this 'top and key technology' and that future benefits, including economic benefits, should be balanced against ethical responsibilities (ibid.). In a newspaper interview Schröder stated that economic goals were indeed 'ethical goals' (Schröder 2001, 95). Schröder considered scientific 'progress' to be not only inevitable but necessary for the well-being of a liberal and democratic society, and he argued against it being ruled out by legal 'constraints' (Die Woche, Dec 20, 2000). Scientific knowledge and expertise, in his words, were concerned with liberating humankind from natural or traditional constraints and were the basis for democratic public action (ibid.). He explicitly opposed general restrictions for scientific research, which, from his point of view, would contradict the liberal and economic interests of the nation (ibid.): "It is the task of a democratic society to impede an oligarchy of knowledge—not by prohibition but by educating the general public". "No power in the world", he said, "can hinder knowledge" (author's translation, ibid.).

Quite bluntly, Schröder referred to the *imperative of scientific progress*, suggesting that science is inexorably 'progressing'. Against this background, he relegated science policy-making to decision-making on technological applications and effects—because scientific knowledge is continually and inevitably gaining momentum. Schröder decreed that science policy must be constantly renegotiated in the light of new techno-scientific developments and not vice versa. (Political) decision-making, as he defined it, tends to follow scientific 'progress' rather than precede it. Indeed, he declared that any restrictions in the area of scientific development were 'irresponsible' (ibid.) and therefore unacceptable. By doing so, Schröder neutralized the intervention of the public and policymakers in shaping scientific development. In fact, scientific development was excluded from problematisation altogether. In other words: problematisation of unobstructed scientific development was declared to be 'irresponsible', 'ideologically biased', and unrealistic.

Several actors introduced economic arguments into the debate. Advocates of new biomedical developments, such as PGD or embryonic stem cell research, argued that a ban on these practices would seriously damage German competitiveness within the global biotech economy. The key arguments along these lines were that companies, researchers, and users would leave the country and that, if forbidden in Germany, therapeutic benefits deriving from practices would have to be imported and this would seriously damage the German job market. In sum, the scientific and economic future of the nation was presented as being at stake if restrictions on research were implemented or, as it were, sustained. In other words, maintaining or strengthening Germany's economic and scientific position in the world, in the view of these actors, was framed as an 'ethical' goal requiring the deregulation of science (cf. for example Reimann 2000; von Renesse 2000; BT DR 14/4098). For these actors, who were committed to a belief in 'societal progress through scientific progress', the challenge was not to contain knowledge production and restrict or shape research practice, but to set the foundation for its unconstrained development; their goal was not to prohibit or control research but instead to establish public acceptance of it. A public debate, so the argument went, should help create acceptance of scientific development—excluding, at the outset, a debate on whether or not scientific innovations should be restricted (Frankfurter Allgemeine Zeitung, Dec 12, 2000)

While economical arguments were not in fact entirely new, what is noteworthy in Schröder's liberal argumentation is the fact that economic 'benefits' and economic competitiveness were introduced as high-ranking *ethical* stakes that were balanced against any other (problematic) moral implications or 'dangers'. From this perspective, scientific progress was presented as a prerequisite for the well-being of the nation and therefore excluded from problematisation. Instead, the 'dangers' for the economic future of the country were interpreted as a problem. However, Schröder's engagement in the debate, especially his plea against "ideological blinkers" or "general restrictions" and the fact that economic arguments were introduced into it, was widely interpreted as a questionable intervention from the top of the rung in the principal course of the debate (cf. Graumann 2002b, 21). In other words, Schröder was accused of trying to use his powerful social resources as chancellor in order to trigger and frame discourse in economic terms. However, although the economic problematisation came up repeatedly and was put forward particularly bluntly in Schröder's speeches, it was rejected. In fact, even official actors, including techno-optimists, stated that economic considerations should be disregarded where ethical or constitutional objections to human embryo research exist (German National Ethics Council 2001, 12). And furthermore, as the Enquête Commission on "Law and Ethics of Modern Medicine" (EK REM) stated, economic goals should, in general, be viewed as subordinate to the criteria of the right to protection of the embryo and to the high-ranking goals of therapy and research (Deutscher Bundestag 2001, 52).

'Status of the Embryo'

A few weeks later, Schröder's pro-research intervention was backed by the then designated Federal Minister of Cultural Affairs (*Kulturstaatsminister*), the philosopher Julian Nida-Rümelin, who published a newspaper article in which he declared that embryos were not subject to human dignity because they lack self-esteem (Der Tagesspiegel, Jan 3, 2001). Nida-Rümelin's intervention was the starting point for a second frame change within the debate—triggering a semantic narrowing down of the discussion, which now focused on the *moral status of the embryo,* and, at a more general level, led to a discussion on the validity and scope of the constitutional principle of human dignity. In his article, Nida-Rümelin argued:

> "The respect for human dignity is appropriate under the condition that the dignity of a human being can be violated, that his self-esteem can be removed. Therefore, the criterion for human dignity cannot be extended to embryos. The self-esteem of human embryos cannot be violated" (author's translation, ibid.).

Nida-Rümelin's article provoked an outcry in the German public and prompted several national newspapers to initiate expert debates on whether or not an embryo is subject to human dignity and has a right to protection. One newspaper (Rheinischer Merkur, Jan 19, 2001) pointedly declared in article titles that "The Controversy has been Opened: What is a Human Being?" and "We Fight for Dignity!" (author's translation). Several authors shared Nida-Rümelin's view on human dignity and declared that 'the embryo had no right to life or to human dignity' within the German legal order (cf. e.g. Merkel 2001). "Entitlement" to human dignity, as one actor stated, must be based on the "capacity of free self-determination that is governed by reason" (author's translation, ibid., 56), abilities the embryo certainly does not have.

Other authors were outraged that an incumbent minister publicly annulled the principle of human dignity—not only for embryos but principally for all human beings unequipped with self-esteem (Spaemann 2001, 47). Critical replies emphasised the indivisibility of human dignity and especially that it could not be linked to particular qualities of the individual. These authors stated that the introduction of qualities necessary to 'gain' human dignity would annul this principle on the whole by inevitably ushering in the category of 'life not worth living' (ibid.). A conservative MP and member of the *Bündnis für Menschenwürde* (cf. 4.3) stated that it was "extremely dangerous" to declare self-esteem to be a criterion for human dignity as this would also exclude coma patients or seriously disabled people (Focus, Jan 15, 2001). Thus, living up to its subheading "Germany Cannot Avoid a New Bioethics Debate" (author's translation, Der Tagesspiegel, Jan 3, 2001), Nida-Rümelin's article not only stimulated a discourse but, at the same time, promoted a framing of embryo research or PGD as issues concerning *the status of the embryo*. The intervention introduced a problematisation that centred on issues of whether or not the embryo is subject to human dignity and whether or not it has the right to life. Framing the issue in this way meant that actors on either side had to articulate their arguments in terms of the status of the embryo if they wanted to participate in the debate. At a more general level, the debate ensuing Schröder's and especially Nida-Rümelin's interventions was a debate on the foundations of the German legal order, i.e. the constitutional principles. To be more precise, it was a debate in which the universal idea of indivisible human rights and human dignity were in question. In contrast to discourses connected to the anti-bioethics alliance (cf. 4.3), the argument for 'human dignity' and the 'right to life' was not put forward by techno-sceptical actors to underpin the call for restricting science. In fact, it was introduced by supporters of the new research, who, by doing so, opened up a discursive space for balancing constitutional principles against techno-scientific innovations.

The 'Abortion-Trap'

Framing of PGD and embryonic stem cell research in terms of 'the status of the embryo' led some to relate these practices to abortion practice and law. A preferred argument of pro-research actors emphasised what was called a 'value contradiction' (*Wertungswiderspruch*) as regards 'differences in protection' of the embryo in vitro and in vivo (cf. e.g. Merkel 2001). In one instance this 'contradiction' was called "the abortion trap" (Die Zeit, June 13, 2001). In fact, actors applied what can be called the *abortion trap frame*, arguing that those who are against embryo research or PGD must also oppose abortion and vice versa. One commentator wrote,

> "it is against all reason to accept the killing of 134,000 embryos after nidation [...] and declare, at the same time, that the strictly controlled use of several hundreds of not implanted embryos for high-ranking research purposes would cross the Rubicon" (author's translation, Peter Glotz, quoted in Graumann 2002b, 24).

Abortion practice was also referred to by techno-optimist actors in another sense—especially with regard to PGD: Pre-implantation genetic diagnosis would help avoid abortion and was therefore in the interests of women. This argument had been brought forward already at the beginning of the PGD debates, by, for example, the GfH (1995) or the BÄK (2001). It was, however, not only factually wrong but also legally misleading: Also downright supporters of the legalisation of PGD usually admit that it has to be followed by a prenatal diagnosis (PND) in order to confirm the results (cf. e.g. GfH 1995) and hence does not necessarily exclude abortion. Furthermore, the argument conceals that German abortion law does not allow abortion for embryo-pathological reasons, despite the fact that prenatal tests are regularly performed and often followed by abortion in case of positive test results in Germany. Also the 'avoid abortion' argument is certainly based on the dubious assumption that embryo-pathological reasons will or should even inevitably lead to abortion. Nonetheless, it was supposed to "set the pace" (author's translation, Frommel 2001, 74) for the liberalisation of embryo protection in Germany for techno-optimist actors linking the debates on PGD or embryonic stem cell research with the 'abortion-question'. As the liberal feminist Monika Frommel stated, the debate on abortion "has shown that, even though especially a differentiated liberal viewpoint can emphasise the idea of life protection, it can lead to a reduction of penological prohibitions and thereby to an acceptable balancing of interests" (ibid., 74). Quite clearly, the 'abortion trap' frame was devised to underpin calls for the deregulation of scientific enquiry.

Feminist Interventions

In contrast, techno-sceptical and especially feminist techno-sceptical actors argued that abortion and PGD or embryo research were incomparable practices. While they rejected "concentrating on the 'protection of life' in the public debate" (discussion protocol in ReproKult 2002a, 120), getting involved in the legal debates on embryo protection was nevertheless seen as being based on strategic reasons (ibid.). Especially because abortion was a particularly important topic in the feminist movement, feminists could hardly refrain from intervening in the debates. The main techno-sceptical feminist argument in this respect was that pregnancy was a special situation that was incomparable, also in legal terms, with any other because the embryo only existed in and by means of the pregnant woman (cf. Braun 2003). "It is this real, physical connection that makes differentiating regulations imperative" (Mildenberger 2002). In this respect, the main argument was that an *unwanted* pregnancy—i.e. the main reason for abortions—was a *conflict* situation requiring a solution. Forcing women to give birth violated their right to self-determination and physical integrity, and it is this personal conflict situation that German abortion law acknowledges (ibid.; Braun 2003). Because the embryo in a petri dish exists outside the woman and lacks the real protection of the woman and because artificial reproduction is deliberate and *intended* to create a pregnancy, a *conflict* situation comparable to an unwanted pregnancy does not exist (Mildenberger 2002). Techno-sceptical feminists argued further that the comparison of in vitro and in vivo embryos meant equating pregnant women with 'test-tubes' (e.g. Freitag 23, June 2, 2000; Frankfurter Rundschau, May 30, 2000; Braun 2003).

At another level, feminist techno-sceptical actors not only doubted the 'high-ranking' goals of the research, but they also rejected the 'interests of science' becoming the sole measure for political decision-making, particularly in cases where women's basic rights were at stake (ReproKult 2002b). Permitting embryo research would turn women into a site for conflicts of interest and transform them into a resource for research material that serves also economical interests:

"Embryo research instrumentalises women and men. Their reproductive body substances (egg cells, sperm, embryos) are used as a means to an end in research and in new biomedical markets, or are even to be produced specifically for the purpose of exploitation. The decisive point is that it is not just a question of the intention of enabling a medically/technically controllable pregnancy or making it standard, but also of the availability of embryos as research material, as future 'spare parts', or as economically exploitable and patentable goods. Exploitation outside the sphere of reproduction would implicitly degrade women and men to the status of raw material suppliers for research. The physical integrity of women, in particular, would be violated by the invasive procedures required to obtain egg cells, and women would be demoted to the status of contract suppliers of egg cells or 'surplus' IVF embryos (i.e. intentional overproduction). This is unethical and socially unacceptable" (ReproKult 2002b, 138).

In sum, feminist techno-sceptics actively tried to stop framing in terms of the status of the embryo used to link PGD or embryo research to the abortion question. Instead, they problematised the issues in terms of the social and political consequences, particularly for women, of the new technologies. For them it was crucial to focus on the social consequences of IVF, PND, PGD, and embryonic stem cell research. To this end, feminists emphasised that the feminist concept of self-determination must be elaborated and take all women into account, including those who were disabled or had disabled children. A "context-based definition of 'female self-determination' must be elaborated, which imposes responsibility on women as self-determined individuals, but simultaneously takes into account the social constraints under which women make decisions" (discussion protocol in ReproKult 2002a, 120f). The feminist criticism of IVF, PND, PGD, and embryonic stem cell research "can be soundly justified by the negative consequences for the individual women involved and for the standing of women (with and without disabilities) in society" (ibid.).

Feminists were well aware of the discursive pressure to participate in the debates put on techno-sceptics by emphasising the abortion-trap frame and focusing on the status of the embryo. While they could not avoid getting *strategically* involved in the 'embryo-debates', they nonetheless tried to also introduce a broader problematisation that included social conditions and the consequences of the development and application of new scientific developments. In this respect, the feminist conference "Reproductive Medicine and Genetic Engineering: Women between Self-Determination and Social Standardisation" (ReproKult 2002a) was a decisive event. The organisers sought "to initiate a discussion on aspects of women's policy that receive virtually no consideration in the current public debate surrounding 'biopolicy' and, simultaneously, to develop political strategies for coping with associated technologies" (Herrmann/Kurmann 2002, 6).

Summary: Discourse Explosion

At the end of 2000, we find considerably more discourse in Germany on reprogenetic practices and, in particular, on embryonic stem cell research, triggered by actors promoting scientific developments and/or representing research interests. However, as compared to the years before, we also find a profound transformation of problematisations. Introduced, above all, by Chancellor Gerhard Schröder, who engaged in the debate for the first time, as well as the designated Minister of Cultural Affairs, Julian Nida-Rümelin, problematisations now focused on the status of the embryo, allowing and promoting the comparison of reprogenetic practices with abortion. While abortion already served as a reference point for pro-PGD arguments at the beginning of the debates comparing PGD and abortion, at the time, was used to frame PGD as a kind of early abortion. At the turn of the millennium, in the eyes of pro-research actors the reference point for the abortion trap frame became the 'status of the embryo', serving the rationalisation of the destruction of embryos (in PGD or embryo research). Feminist actors felt that it was strategically essential to engage in this debate, although they did not equate abortion with embryo destruction in research or PGD. They wished to affirm problematisation of reprogenetics in socio-political terms, referring to social pressure on women and rejecting the instrumentalisation of women's needs in favour of scientific endeavour. Feminists and other techno-sceptical actors defended the status quo and wanted to foreclose the deregulation of the Embryo Protection Act.

Discussion on and, indeed, an attack on the fundamental principle of 'human dignity' also entered into the debate. Techno-optimist actors stimulated a debate on the human dignity of the embryo, demanding that also opponents engage in the debate and, most importantly, frame their arguments in terms of the status of the embryo. Although human dignity was the most important reference point for the anti-bioethics alliance in the late 1980s and early 1990s, it was not a discussable principle at the time: for the alliance and for techno-sceptical actors more generally, human dignity was the stable and undeniable principle by which new practices were to be judged, and not vice versa. Instead, the debate on human dignity ignited by Nida-Rümelin demanded of such actors that they prove the applicability of the principle to human embryos and enter a debate they had hitherto rejected.

4.6 Import Stem Cells Now! More Pressures

Pressure on policymakers to come to a decision on the issue of embryonic stem cell imports and research increased in mid-2001, when two researchers from the University of Bonn announced that they wanted to import embryonic stem cells from Haifa, Israel. They were supported by the prime minister of *Nordrhein-Westpfahlen*[104] and backed by the German Research Foundation, who published their "Recommendations for Research on Embryonic Stem Cells" (author's translation, DFG 2001), stating their intention to support and fund research on imported stem cells. Against a background of imports not regulated by law, policy-makers were put under great pressure to speedily make a decision before the DFG and Bonn researchers simply created facts by importing stem cell lines. The government coalition asked the DFG to postpone a resolution on funding until the German *Bundestag* had come to a decision on the matter. A motion by the SPD and the Greens asked both researchers and the DFG to not "anticipate a resolution by creating consummated facts" (author's translation, BT DR 14/6551), that is, by simply importing ES cells. Parliament declared that it would have to wait for the statements of the recently established advisory boards (the NER and the EK REM will be discussed below) before it could come to a decision (ibid.). The deadline for a decision on funding to be made by the DFG, initially scheduled for July 2001, was renegotiated and postponed until the end of the year and deferred again in December until parliament arrived at a decision in January 2002.

The DFG's recommendations on embryonic stem cell research (DFG 2001) again led to an intensified debate in the German press and public. The debates were marked by references to the 'lack' of restrictive legislation in other countries and by the (negative) effects of restrictive legislation in Germany, impacting its competitiveness in biotechnology. They were put forward by techno-optimist actors in order to strengthen their call for the deregulation of embryo protection. Their arguments were framed in terms of the *imperative of scientific progress* as being interrelated with what can be called a *German Sondermoral* frame (German special morals frame) in the case of the status of the embryo. This debate was an extrapolation of the discussions on 'the status of the embryo' and 'human dignity', which were presented above.

104 North Rhine-Westphalia is the main biotechnology region in Germany; about one quarter of German biotech business is located in the region.

The Imperative of Scientific 'Progress' and the 'German Sondermoral'

In the DFG's recommendations on ESC research, it claimed "significant progress in embryonic [...] stem cell research" (author's translation, DFG 2001, 108) since the publication of its last statement on the issue (DFG 1999). While the first statement abstained from calls for immediate changes in the EschG, in its new statement the DFG declared that it was now time for German law be adjusted to scientific 'progress': "The DFG believes that science has now reached a state in which potential patients as well as scientists in Germany should no longer be excluded from these developments" (author's translation, DFG 2001, 109).

Hinting at a legal loophole, the DFG did not consider imports of these cells to be a violation of German law, even though their production was still prohibited in Germany. In fact, the EschG does not cover the *use* of pluripotent embryonic stem cells, but only prohibits the destruction of embryos (necessary for the derivation of ES cells) as well as the use of so-called totipotent cells. Nevertheless, it was basically undecided if other legal restrictions applied, such as Art. 7 Sec. 2 of the German penal code (StGB). Some actors interpreted this article as specifying that German imports of ES cells not be related to their derivation and that German actors not be involved in the destruction of embryos (cf. Deutscher Bundestag 2001).

Against the background of these legal uncertainties, the DFG pleaded for a harmonious adjustment of German law, framing the German legal situation in terms of what could be called a *German Sondermoral*, that is, a 'special ruling' for Germany in comparison to other countries:

> "Regarding this matter, if we accept that legal differences in international comparison are not per se objectionable and that, excepting any injustice outlawed throughout the world, activities in a foreign country have to be evaluated in relation to what legally applies there, then it is not justifiable to principally rule out research on embryonic stem cells created abroad if we take the constitutional guarantee for the freedom of research into account" (author's translation, DFG 2001, 109).

The *German Sondermoral framing* was substantiated against the background of the assumed *imperative of scientific progress* by the then president of the DFG Ernst-Ludwig Winnacker in a speech held at the annual meeting of the DFG in July 2001. A prohibition of embryonic stem cell research in Germany would pose, as Winnacker stated, a "dilemma" for science and society: Science in Germany, so he asserted, would be deprived of the instruments to "prove" that "this research is inevitable" (author's translation, Winnacker 2001, 120). At the same time, society "simply and silently lets the risks of research be taken abroad" (author's translation, ibid.). In order to avoid "double standards", that is, prohibit research in Germany while using scientific achievements from abroad, the DFG proposed that German researchers should themselves "apply for permission to create embryonic stem cells" (author's translation, ibid.).

The DFG framed 'scientific progress' as *inevitably* continuing (somewhere in the world, even if not at home) and explicitly excluded political or societal intervention in ongoing scientific research. At the same time, *society* and *not* science was declared to be responsible for (the displacement of) risks of scientific research on the grounds of an alleged German *Sondermoral*. Winnacker suggested that the simplest answer to this situation was to allow the research. Along these lines, Chancellor Schröder, speaking at the same conference, argued that "nobody can simply stop scientific progress, at least not on a national level" (author's translation, Schröder, July 5, 2001). In sum, the DFG—backed by Gerhard Schröder—interpreted the issue of embryonic stem cell imports and research as a conflict between the opposing poles of the imperative of scientific progress and a German *Sondermoral*. What was declared to be the German *Sondermoral* was, in fact, the constitutional right to life. Therefore, it is not surprising that the DFG suggested the conflict be solved by balancing two constitutional rights: the right to life and the freedom of research (DFG 2001, 111). The outcome of the balancing process, however, was already set: "the DFG holds that the *Rubicon had already been crossed* by introducing artificial fertilisation and that it would be unrealistic to believe that, surrounded by decisions already made on the right to life of the embryo ([...] introduction of nidation blockers, abortion), our society can return to the previous status quo" (author's translation, ibid., 114, emphasis added). Scientific development, from the viewpoint of the DFG, had already, once and for all, breached the potential limits. Thus, anticipating that the balancing process would end in favour of the 'freedom of research', the DFG proposed guidelines in the same paper for the derivation of as well as research on embryonic stem cells (ibid., 112ff).

The DFG's statement and, in particular, the emphasis on a 'legal loophole' within the EschG put the question of importing embryonic stem cells for research at the top of the political agenda. In the run-up as well as aftermath of the first great biopolitics debate in the German *Bundestag* (PLP

14/173) scheduled for the end of May 2001, a debate arose in the press and among the public centring on the question of whether it was morally acceptable or not to import cells derived from the so-called surplus embryos, although the production of these cells was prohibited in Germany, or if it was reasonable to change the EschG and to legalise the production of surplus embryos and embryonic stem cells in Germany. Embryonic stem cell research and the import of stem cell lines became politically the most pressing and conflict-ridden issues in German biomedicine policy. The national newspaper *Frankfurter Allgemeine Zeitung* (9 May 2001) spoke of a "heated autumn" of biopolitics, in which far-reaching decisions had to be made, comparing the conflict potential of stem cell research with the sometimes violent protests surrounding the deployment of American Pershing II missiles on German territory in the 1980s.[105] The discussion on the scope and validity of human dignity as a case of, as it were, German *Sondermoral* and the relation between human dignity and the governance of science were at the top of the scale. Indeed, the problem was henceforth constructed as a conflict between a German *Sondermoral* and the freedom of research, with particular consequences for suggestions of how to solve this conflict and how to govern scientific research.

A Case of German Sondermoral: Human Dignity

In May 2001, the then Federal President Johannes Rau held a widely acclaimed speech in Berlin in which he emphasised the need of boundaries for science and research without necessarily risking setbacks in medical treatment. Explicitly rejecting the imperative of scientific progress and pleading for common boundaries for scientific development instead, Rau said that "there is a lot of room on this side of the Rubicon" (author's translation, Rau 2001, 18). Rau emphasised that science and research can accomplish good without "entering ethically critical areas" (author's translation, ibid.). He also emphasised the idea of human dignity and opposed its association with the qualities brought forward by Nida-Rümelin (Der Tagesspiegel, Jan 3, 2001) in January. All in all, Rau proposed that technoscientific development should be judged against the background of basic societal values and the impact science and technology have on them—including also the possibility of stopping or, at least, slowing down scien-

105 The term "Heißer Herbst" was used in Germany in the 1980s to refer to the protests against Pershing II deployment.

tific development and application (Rau 2001, 27). For this view the idea of human dignity was the key principle. Rau held human dignity to be the only acceptable and indisputable touchstone for arriving at mutual decisions concerning scientific developments (ibid., 19).

Other actors, however, interpreted a restriction of research based on the idea of human dignity as an inappropriate and a singularly German argument in the debate, which contradicted the liberal regulation of research in other occidental democracies (e.g. Markl 2001). The then president of the Max Planck Society Hubert Markl stated in a speech that Germany should "balance the arguments of other occidental nations founded on constitutions no less free or democratic than ours before we attempt to occupy the lofty shores of ultimate moral justification together with the Vatican" (author's translation, ibid., 184). Arguments within the *Sondermoral* frame were directed against the idea of human dignity being absolute. Markl suggested that the freedom of conscience and action be an alternative focus. However, on closer inspection, this turned out to be mere freedom of choice (ibid., 184): In Markl's terms, what makes a human being a human being is "his culture-generating freedom to decide" (author's translation, ibid., 189).

Applying a German *Sondermoral* frame was additionally directed against the idea of arriving at mutual political decisions on the development of science and research. Indeed, within the *Sondermoral* frame the 'freedom to decide' explicitly excluded the possibility or even legitimacy of political action and collective decisions in science policy in favour of individual choice according to personal morals and conscience. As Markl stated: "There can be no binding opinions concerning these questions. [...] Instead I have to individually account for them with all my error-prone convictions" (author's translation, ibid., 187). The fact that Markl spoke explicitly not as the president of the *Max Planck Gesellschaft* but as "citizen Markl" (ibid., 178) underscored that no authority—neither the president of the republic (Rau) nor himself, an ethics council (the NER or the EK REM), nor the legislator—can ever come to a collective decision on these issues. Instead, Markl offered a specifically liberal 'solution', framing the issue as a matter of personal moral convictions:

> "Because, in questions of conscience, morals, freedom, responsibility, and human dignity, all citizens must be their own experts and avail themselves of the freedom of speech to the best of their knowledge and in keeping with their conscience" (author's translation, ibid., 178).

Problematisation of a German *Sondermoral* and the presentation of a solution in terms of personal moral conviction were based on the belief that scientific progress was not only inevitable but, indeed, necessary for the existence of a society. Scientific progress was considered to be imperative and thus beyond problematisation and political intervention: Freedom of research, Markl maintained, is the prerequisite for the "existence of culture and the state itself" and not a "political gift from the state that was graciously bestowed on scientists" (author's translation, ibid., 190).

Framing the issue of embryonic stem cell research in this way was criticised as introducing a strategic "trial balloon" (Schirrmacher 2001) to underscore accusations of fundamentalism on the part of those who voiced more general doubts against ES cell research (Sahm 2001, 200). Indeed, the understanding of the research as a matter of 'the freedom to decide' was attacked as too simplified and some actors argued that the issue was not about the "freedom to decide" but the "rightness" of the practices in question (author's translation, ibid., 201). Statements referring to other countries with more permissive regulations on the research and statements backing the *Sondermoral* framing were criticised because they ignored the fact that disputes over the pros and cons of embryo research or other bioscientific or biomedical practices currently existed also elsewhere. Additionally it was asserted that there were, in fact, countries other than Germany with a restrictive policy on these issues (ibid.).

Summary: Human Dignity and 'Progress' of Science

When the issue of embryonic stem cell imports and research reached the top of the political agenda in Germany at the beginning of 2001, an overall framing of the issue (as well as of scientific research more generally) became dominant in which scientific 'progress' and 'moral'—i.e. constitutional—concerns had to be balanced. The antithetical references to a so-called German *Sondermoral* and the implicit or explicit plea for harmonising different national regulations was meant to underpin calls for the deregulation of ES cell research. Framing embryonic stem cell imports in terms of a German *Sondermoral* substantiated problematisation of a conflict between (allegedly illegitimately) adhering to a 'German singularity' of human dignity versus freedom of the individual, freedom of science, and development of society. Speaking more generally, the German *Sondermoral* frame helped reinforce problematisation of collective decisions and restrictions regarding science—even if they are based on the constitution—as an illegitimate limitation of freedom and of the well-being of society. In other words, this framing was devoted to the call for ongoing scientific research, which was to proceed without obstruction and outside of political decision-making processes as a prerequisite for individual freedom and societal development. Therefore, the *imperative of scientific progress* underlies the overall framing also for the understanding and interpretation of 'human dignity' or the 'status of the embryo'. Within it, any wish to control or restrict science, even on the grounds of the constitution, was interpreted as a case of German *Sondermoral*. Although techno-sceptical actors, foremost the then Federal President Johannes Rau, criticised these arguments and the new techno-scientific practices as attacks on the principle of human dignity, the overall framing of the issue of embryonic stem cell research (and scientific research in general) as a question of balancing scientific 'progress' and 'moral' (constitutional) concerns became dominant. The *Sondermoral frame* persisted and the slogan "research from abroad—ethics from Germany" (author's translation, Berliner Zeitung, May 27, 2001) became a popular saying in liberal pro-research circles. Inherently, this slogan implied that research continues and will always do so somewhere in the world, excluding, in principle, any political—especially restrictive—intervention and the possibility of *not* using or restricting the use of research results in Germany.

4.7 A Speaking Cure for Conflicts? The Institutionalisation of Discourse

In Germany, the years 2000 and 2001 were not only marked by more public but also more institutionalised discourses. Within this period several commissions or processes connected to official politics were established: Health Minister Andrea Fischer's Symposium on Reproductive Medicine (2000), the parliamentary Enquête Commission on 'Law and Ethics of Modern Medicine' (2000; EK REM), as well as the German National Ethics Council (NER), which was set up by Chancellor Gerhard Schröder in 2001. In Germany, discourse on PGD or embryo research was institutionalised much later than in the UK. Whereas in the UK high-ranking policy-makers and commissions were involved in and led the debate from the very beginning even before conflict arose, the institutionalisation of discourse in Germany was (partly at least) a result of public conflicts surrounding bioscientific developments. What was, in the UK, a precautionary measure to avoid conflict, was, in Germany, not only the result of conflict but likewise a 'solution' to it. This 'solution', however, was a device that was implemented to structure and (re-)organise public debate. In this respect, it was comparable to institutionalised discourse in the UK. The established institutions and processes were meant to be an answer to the great public conflicts and prepare legislative or parliamentary initiatives concerning reproductive medicine and ES cell imports and research. These initiatives also reveal the transformation and narrowing down of problematisations within the public debates discussed above, because they attest to emerging problematisation of discourse practice that led to calls for a *new* discourse as a solution.

A Symposium on Reproductive Medicine

In the great public conflicts at the end of the 1990s on PGD, in which the question of whether or not the Embryo Protection Act prohibited genetic testing of in-vitro embryos played a prominent role, the then Health Minister Andrea Fischer was under great pressure to draft a new law on reproductive medicine (FmedG) and to create legal certainty. The draft guidelines on PGD, which were published in March 2000 by the BÄK (2001) and discussed above, generated much protest among the German public. Many suspected that the BÄK wanted to pre-empt the drafting of a new act on reproductive medicine, which had already been announced by the minister of health. Fischer had set up a ministerial ethics board (*Ethik-Beirat*) to prepare the drafting (Interview D4, 10 Nov. 2005).[106] Fischer likewise announced that a symposium on "Reproductive Medicine in Germany" (BMG 2001) was to be organised, which was also explicitly directed at the drafting of a FmedG. While the appointment of a ministry expert commission is a common device in the process of preparing policy-mak-

106 Fischer set up an Ethics Board within her ministry to prepare the drafting of a new act. The only topic the Ethik-Beirat dealt with was genetic testing (Ethik-Beirat 2000). Reproductive medicine was postponed to a later date, but the board was suspended before it could discuss the issue (Interview D7, Jan 18, 2006). Since 1998 (at the latest), the issue of genetic testing was on the political agenda, when the Social Democrats and the Greens declared in their coalition contract that they intended to implement regulations against genetic discrimination. An act on genetic testing was repeatedly on the agenda, but, still today (Feb 2009), has not been passed, despite the fact that there have been several initiatives and announcements to publish a draft and although policy-makers agree unanimously on the necessity of such an act (Riedel 2004).

ing, in order to prepare drafting, it is rather unusual that a deliberation process as extensive as Fischer's symposium be organised.

The symposium,[107] which took place in May 2000, was exceptional in many ways: in terms of size and budget (there were about 600 participants, who contributed to the discussions with almost 70 speakers for a period of three days; cf. list of speakers in BMG 2001, 514-519) and in terms of the enormous media coverage it received. Speakers and participants came from very different fields, including the life-sciences, reproductive medicine, social-sciences, ethics, social-work, midwifery, as well as from the disability and the feminist movements, NGOs etc. The combination of experts and different actors attending was regarded as quite exceptional. Not least because of the number of feminists (certainly of heterogeneous kind) involved and because of the fact that panels and individual speakers comprised many actors who were sceptical towards the life-scientific developments, especially PGD and embryo research. For these reasons, the conference was called a "step towards a new political culture" (author's translation, Berg et al. 2001).

The symposium was rather exceptional also at a semantic level in that it encompassed a wide range of issues and perspectives that went beyond merely scientific accounts. The programme of the event (cf. table of contents in BMG 2001) reflected the spectrum of views on and problematisations of scientific developments existing in the German public: Not only did many of the key issues focus on a wider range of socio-political conditions and consequences of bioscientific development as well as on science-based accounts. Also a constituency of invited experts represented problematisations that went beyond a science-focused view on the issues at hand. The Fischer symposium abstained in many ways from a common *first get the facts right* format of discussion on biomedical/bioethical issues, in which usually a discussion of 'ethical' and (if at all) of social and political aspects follows scientific explanation or reasoning. Fischer's wish to avoid a scientific expert bias was mirrored not only in the variety of participants but also in the guiding questions for presentations and discussions.

However, Fischer, in organising the symposium, was only partly motivated by the need to obtain advice on how to deal with issues or by the wish of making a topic public that she had hitherto considered a purely expert issue (Interview D4, Nov 10, 2005). The aim of the conference was to

107 The author attended the symposium.

"prepare a process" not only in terms of "content" but likewise in terms of "legitimation" (author's translation, ibid.). Additionally, it was to be a remedy for what Fischer called "an extremely charged bioethics debate" (author's translation, ibid.). In her view, the debate on reproductive medicine was bogged down, and a consensus or even compromise seemed unattainable. Fischer sought to bring a lot of very different groups together to "speak to each other" and to integrate the whole range of "relevant viewpoints" in the discussion. The motivation was to give the societal groups involved a "fair chance to be integrated already in the thinking process" (author's translation, ibid.) prior to drafting the bill. Therefore, apart from substantial conflicts, Fischer was also concerned about conflicts over discursive practices and conflicts over the forms and styles of interaction present in the ongoing 'bioethics' debates—problems that the symposium was meant to resolve. As Fischer declared: "The terrain was mined. [...] The symposium was based on the somewhat naïve hope to break up something completely" (author's translation, ibid.).

As a remedy for conflict in this field, Fischer suggested a certain style and form of argumentation that was meant to tame the adversarial elements in the debate: "Such a debate can only take place if one respects the opinions of others, without thinking [...] that the others have barbarian views" (author's translation, Fischer 2001a, 329). 'Respect' for the opinion of others signified that supporters of the new technologies should not denounce opponents as 'fundamentalists' driven by 'irrational fears', whereas opponents were not to accuse supporters of being motivated by 'economic interests'. Thereby it was implicitly declared that the debate was free or could be freed from power relations and particular interests. By this means, fundamental differences between the participants and adversarial perspectives could not be addressed, especially if they referred to power relations and interests.[108] In sum, in Fischer's symposium we find the incentive to transform an adversarial political debate into a debate on various, equally valid, and respectable 'opinions' held by 'uninterested' actors. The actors likewise had no claim to holding the 'right view' on the issues at hand, but were, in principle, to accept all the different perspectives—even the totally antagonistic ones—as 'equally respectable' and worthy of discussion.

108 In fact, the (sparse) comments referring to economic or research interests met with aggressive reactions.

Furthermore, Fischer was well aware of the conflicts concerning science governance and she was quite explicit about what she thought was the best way to deal with these issues. She insisted on the need for *common rules* and not for professional or individual rules "without obligation to homogenous values" (author's translation, ibid., 329). In her concluding speech at the symposium, Fischer explicitly declared that, from now on, the matter was the responsibility of politics and *not* that of science and expertise or of the individual.

> "All speakers participating have declared, in rare unanimity and modesty, that their disciplines alone are not in the position to make the necessary decisions. Thus, we have to think beyond the borders of the respective disciplines, and this tosses the ball back to politics" (author's translation, ibid., 327).

Consequently, a few months later in November 2000, Fischer published a position paper (BMG 312-4080/17) outlining the core points of a new act on reproductive medicine. The act was explicitly supposed to replace the EschG without, however, diminishing embryo protection. On the contrary, the new act was to entail the high standards of embryo protection in Germany while regulating new developments in reproductive medicine. The position paper suggested a ban on PGD on the grounds that the definition of indications for PGD would require a morally questionable selection of illnesses interpreted as 'sufficiently' serious to indicate PGD (ibid., 4). Fischer also strongly opposed the analogy between PGD and abortion, which especially supporters of PGD had put forward against its prohibition. Taking up an explicitly feminist perspective, Fischer argued that permitting PGD and the consequent selection of embryos after a genetic test would negate the 1995 removal of the 'embryo-pathological indication' from German abortion law (ibid., 5). The paper also promoted a ban on egg cell donation and on the derivation of embryonic stem cells. The import of these cells was to remain allowed as long as they were pluripotent. Research in which an embryo is destroyed as well as cloning and germ line intervention were also to remain prohibited. While Fischer always had a clearly sceptical view of the new reproductive methods, she still was certain that "in bioethical questions it is important to find a compromise" (author's translation, Interview D4, Nov 10, 2005). The paper should "present a stance" and provoke discussion also within parliament (ibid.). However, as Fischer explained, a discussion in parliament did not take place because the political pressure to make a law had abated (ibid.).[109] Fischer could not bring the drafting process to an end, as in January 2001 she resigned from office because of accusations of mishandling BSE in Germany. [110] The health ministry was handed over to Ulla Schmidt (SPD), who immediately cancelled the project of drafting, let alone implementing a new FmedG. The Ethics Board within the BMG was suspended and it was proposed that a National Ethics Council be established. A FmedG, however, has been put on hold since then. As Fischer said in the interview, the symposium, in retrospect, was nothing but a "straw fire" (ibid.).

A Parliamentary 'Struggle Commission'

A second official initiative was establishing the parliamentary Enquête Commission on "Law and Ethics of Modern Medicine" (EK REM)[111] in 2000. The establishment of the EK REM was a result of demands by the protest movement against the European Bioethics Convention (cf. 4.3). Although public pressure was an explicit reason for setting up the commission, doing so was not an easy process (cf. BT DR 14/3011). Despite the fact that even many MPs from all parties supported that a biomedicine enquête be established, it was defeated by parliament in 1999. Reasons for rejecting the commission were diverse. However, an important point was

109 It is conceivable that political pressure was now against implementing a new act. This situation can be compared to that of the 1980s, when doctors and scientists successfully prevented the implementation of a nationwide act on reproductive technologies.

110 In November 2000, Germany had its first BSE case. Although stipulating that beef be tested, Federal Health Minister Fischer was said to have misjudged the extent and the health risks of the increasing number of BSE cases. On 9 January 2001, Fischer as well as the Federal Minister of Agriculture resigned, as both were accused of mismanagement in the case of BSE.

111 Enquête Commissions are a common parliamentary device to investigate certain issues in more detail before parliament makes a decision on them. They are usually set up for the duration of a legislative period and for a certain topic. The conclusions made by the commissions are the basis for parliamentary decisions but are not binding (cf. GoBT Art. 56). Formally, the parliament has to set up an Enquête Commission if one quarter of the MPs supports a motion. According to parliamentary standing orders, the motion must name the topic as well as outline the problem. Although the parliamentary standing orders state that each party can delegate one party member, usually the number of party members chosen by parliament for such a commission is proportionate to that in parliament. Additionally, a number of experts are appointed by the president of the Parliament or by the parties themselves. All commission members must have the same rights (Unkelbach 2001).

that the motion for setting it up refused to sign the European Bioethics Convention (Interview D1, Oct 21, 2005; Catenhusen, June 29, 1999; cf. also Geisler 2002). Also, opponents of the commission asserted that a new commission would postpone necessary legislation (Interview D8, Jan 19, 2006), stating that there had been sufficient debate on the issues (Catenhusen, June 29, 1999).[112] The commission was called "a struggle commission" (*Kampf-Enquête*) and nothing but a "parliamentary arm of the oppositional movement" rather than an "open forum for debate" (author's translation, Interview D1, Oct 21, 2005). As a result of the strong opposition against setting up the EK REM, especially within the government coalition, the motion for establishing it was defeated in parliament leading, however, to a wave of public and inner-parliamentary protest. Citizen initiatives, organisations of disabled people, members of the Church, and MPs campaigned against this decision.[113] In March 2000, parliament finally decided to establish the EK REM for two years, which was supported by all factions (BT DR 14/3011).

The Enquête was supposed to make recommendations for an 'ethical evaluation', a possible societal handling of as well as legislation for biomedical developments (ibid.). It was to take the views of concerned actors, groups, and institutions as well as the Church into consideration. The commission was to present the state of the art of the new technologies as well as ensuing problems, and consider ethical, constitutional, social, legal, and political aspects. Its object was especially to pinpoint legal voids and work out criteria for limiting medical research and application—with special emphasis on the 'preservation of human dignity', as the parliamentary motion prescribed (ibid.). The final report was due within the current legislative period, in March 2002 (ibid.). The EK REM was also to partake in ongoing preparations for new legislation on issues falling within its domain. This was especially crucial regarding the ongoing debate on importing embryonic stem cells. Altogether, the EK REM published two interim reports: on the "Protection of intellectual property in the area of biotechnology" (author's translation, BT DR 14/5157) and on "Stem cell research" (author's translation, Deutscher Bundestag 2001). The final report was published in 2002 (Deutscher Bundestag 2002).[114]

---

112 The letter is not public, but a hard copy was available to the author.

113 Cf. Webpage of the *Tübinger Initiative* at: http://www.fuente.de/bioethik/, Feb 7, 2007.

114 Available also on the EK REM website: http://www.bundestag.de/parlament/gremien/kommissionen/archiv14/medi/medi_ext.html, Oct 10, 2006.

While the process of appointing the members of the commission provoked several conflicts (cf. e.g. Süddeutsche Zeitung, Mar 25/26, 2000),[115] the participation of actors originating from the protest movement against bioethics and biomedicine was highly appreciated by techno-sceptical actors. Establishing the EK REM was not only highly politicised and adversarial in character but was in itself interpreted as a political instrument that, once set up, could be strategically used to strengthen the respective political goals. Especially techno-sceptical actors saw the EK REM as a means of strengthening their position, that is, for making techno-sceptical views heard in public and possibly integrating them in the decision-making process (cf. Südwest Presse Ulm, Feb 3, 2000), even if only because of the prestigious label 'EK REM'.[116] Obviously, the interpretation of the EK REM as a 'struggle commission' was in fact appropriate because members as well as public actors believed that it could be a forum in which adversarial political conflicts that introduce and acknowledge techno-sceptical arguments could take place, and if need be, fought for. The assumed (or hoped for) techno-sceptical political impact of the EK REM becomes even more obvious if one takes a look at another board of ethical advice that was established, the German National Ethics Council (*Nationaler Ethikrat*, NER). In this case, however, the council was formally linked to the government and not parliament. The NER was established by Chancellor Schröder about one year after the EK REM came into existence and quite clearly in order to create a counter-board to it. From the viewpoints of Schröder and other members of government, a *new* discourse was needed that included establishing a new institution.

The Necessity of a New Debate: Setting up a German National Ethics Council

As has been said above, narrowing down the debate semantically to focus on the status of the embryo at the end of 2000 was accompanied by changes in the Health Ministry, which was responsible for and already in the process of preparing a new act on reproductive medicine: When the new Health Minister, Ulla Schmidt, took over the Health Ministry in Janu-

115 Cf. also *Arbeitskreis Bioethik Braunschweig*, webpage, at: http://www.fuente.de/bioethics/bs_000405.htm, May 26, 2005.

116 A similar argument was put forward by a rather sceptical member of the NER, who stated that, by using the name of the NER, she could also confront the public with more sceptical statements (Interview D7, Jan 18, 2006).

ary 2001, she immediately cancelled the drafting of a new act. This change was interpreted as a change of policy in biotechnology towards a "more pragmatic" course (Frankfurter Rundschau, Jan 16, 2001). At the same time, several members of the government, including Chancellor Schröder, declared that they did *not* consider PGD to be forbidden by the EschG and that they would support it in certain cases (Der Tagesspiegel, Jan 25, 2001, Frankfurter Allgemeine Zeitung, Jan 26, 2001).[117] Other members of the cabinet, however, declared that they considered PGD to be prohibited and that embryo protection was secured (Berliner Zeitung, Jan 19, 2001).

Changes in biopolicy were accompanied by the emergence of problematisation of discourse practice: i.e. by calls for a *new* debate including a new style of discourse, new participants, and a new institution. Members of the government disqualified the ongoing discourse as "disordered and unstructured" and too "emotional" (Catenhusen 2001) to substantiate calls for a *new* discourse. Health Minister Schmidt suggested that the chancellor establish a National Ethics Council, which was to be "the place for *responsible discussion* at the executive level" (author's translation, Schmidt, Jan 31, 2001; emphasis added). She stressed that it was important to continue discussions that had recently appeared "in the preceding days" (ibid.), implicitly ignoring the year-long conflicts over and debates on these issues. By emphasising that the NER will be a place of *responsible* discussion, Schmidt in fact declared all previous discussions to be 'irresponsible' and therefore not worthwhile—including, we can assume, also the parliamentary EK REM. Additionally, a new (institutionalised) discourse (the NER) should explicitly *replace* the drafting and implementation of a new act on reproductive medicine. For Schmidt, in regard to PGD or embryonic stem cell research, a new debate was needed before a decision could be made: "If we don't know as a society where to set the ethical and cultural boundaries, new prohibitions don't make any sense" (author's translation, as quoted in Frankfurter Rundschau, Feb 5, 2001). Given the absence of unequivocal legislation on PGD and on embryonic stem cell imports and research, replacing a decision by discourse was, at the same time, a decision for ongoing research and practice without any obstructions—at least as long as (the new) discourse continues. With government authority, the stimulation of a new discourse, so it can be concluded, was to prevent a decision from being made. This "nondecision" (Bachrach/Baratz 1963),

117 At the same time, state funding of biotechnology was more than doubled (cf. BMBF, PM 7/2001).

however, was in fact a decision for ongoing scientific practice of the methods under debate.

Calls for a new discourse were substantiated in February 2001, when the chancellor announced that the German National Ethics Council (*Nationaler Ethikrat*, NER) be set up within the chancellery and installed by himself (die tageszeitung, 13. Feb 2001). The NER was intended to "intensify" the debate, make it "more transparent", and facilitate "informing" the public (Schröder, 5 Apr. 2001). Remembering Schröder's call for a debate "without ideological blinkers" and giving more consideration to economic aspects (author's translation, Die Woche, 20 Dec. 2000), the press and public (for a summary of media responses cf. GID 2001) as well as parliament protested fiercely against establishing the board, not least because this move was interpreted as being directed against the EK REM. In this context, setting up the NER likewise initiated an *alternative* discourse to that of the protest alliance (which resulted in establishing the EK REM) or of techno-sceptical actors. While the EK REM had been interpreted as an instrument of political struggle, the NER was much more generally a "national forum for dialogue".[118] It was an extra-parliamentary expert commission and explicitly *not* meant to be an institution of political struggle. The new discourse was therefore not only intended to replace a decision but also to replace political struggle with institutionalised 'dialogue'.

The discussion about setting up a National Ethics Council appeared at a time when embryonic stem cell research was at the top of the political agenda in Germany and Europe. The British House of Commons had just legalised so-called 'therapeutic' cloning and Tony Blair had announced that Britain sought to become the world's leading biotech nation. The potential economic and medical advantages promised by stem cell research were extensively advertised, in which Chancellor Schröder took a leading role. The first application for funding such research had been filed to the DFG. Many actors critical of setting up the NER, including MPs, complained that Schröder used his authority as chancellor to intervene in the debate in order to achieve the outcome he preferred (Interview D8, Jan 19, 2006). The newspaper *die tageszeitung* (Feb 13, 2001) interpreted Schröder's board to be a means "of offering security" for the extension of state funding of biotech research, implying that the NER was meant to prevent restrictive regulation. Many complained that the issues in question should be subject

118 German National Ethics Council, Draft Decree at: http://www.ethikrat.de/_english/about_us/decree.html, Feb 23, 2006.

to democratic parliamentary and societal rather than expert deliberation and decision-making (e.g. Interview D5, Nov 30, 2005). They stated that the process of setting up the council was undemocratic and that the label 'national' would suggest a representativeness which the council did not in fact have (ibid.; Berliner Morgenpost, May 15, 2001). Especially within parliament, critique of Schröder's move was intense. In the first great parliamentary discussion on 'law and ethics of modern medicine' (PLP 14/173) in May 2001, many parliamentarians, independent of their own positions on PGD or embryonic stem cell research and irrespective of their party affiliation, argued that the NER did not have the democratic authority to make decisions concerning the issues, insisting on the fact that the "*Bundestag* is making the law" (author's translation, Dr. Wolfgang Wodarg (SPD), PLP 14/173, 16906-B).[119] Critics of the NER feared a loss of parliamentary and public influence in favour of technocratic, expert-based, authoritarian government decision-making (cf. e.g. Der Tagesspiegel, May 15, 2001). Schröder replied that the NER will not be an "alternative parliament", but that its task was "to broaden public debate and make it more competent" as well as to "provide competent advice" (author's translation, Gerhard Schröder, PLP 14/173, 16893-B). Schröder emphasised that the public would have to learn more about the scientific details and about the "therapeutic and economic benefits", adding that a "truly public debate" was needed for which the NER should provide the scientific and evaluative grounds (author's translation, Frankfurter Allgemeine Zeitung, Mai 3, 2001). With increased knowledge, society would be able to "openly discuss the options" and decide on the use of new genetics (author's translation, Schröder, Apr 5, 2001). Given Schröder's support of scientific research and against the background of the great pressure put on decision-makers by researchers wanting to import ES cells, his plea for 'more discourse', for a 'true public debate', can only be interpreted as the plea for the stimulation of discourse in order to avert a (restrictive) decision. At the same time, based on a deficit model of the public understanding of science, Schröder was led by the belief that more information about the science would lead to more supportive public attitudes towards it. A public discourse, organised and 'nurtured' by the NER, was supposed to aid in producing acceptance. Altogether, establishing the NER triggered problematisation of not only participation in but also of the forms and processes of

---

119 Cf. also other contributions, for example, Friedrich Merz (CDU), (PLP 14/173,16894-C), Rezzo Schlauch (Bündnis 90/die Grünen), (ibid., 16896-B).

the discourse on biotechnology and biomedicine. This move additionally rekindled controversy over particular practices and contents of the debate. In addition, it was a struggle over the issue of who has or should have the power to intervene in or lead the debate and on what grounds. The setting up of the National Ethics Council and the process of appointing its members was difficult because it was under constant surveillance by the German press and public.[120]

The *new discourse* was institutionalized in the form of the NER in May 2001 (constituted in June). Apart from the provision of statements on 'ethical issues related to new developments in the field of the life sciences',[121] an important task of the new council was to support and initiate public debate. According to the decree, the NER "shall be established as a national forum for dialogue on ethical issues in the life sciences".[122] It "shall be the central organ for interdisciplinary discourse" as well as "organize the social and political debate and ensure that all the relevant groups are involved." Also it had the task of providing 'information material for discussions' and holding public conferences on ethical issues', as well as reporting to the chancellor once a year "on the situation of the social debate".[123]

As a result of the conflicts on setting up the NER and of problematisation of public discourse, a strong emphasis was put on the board's and its members' independence[124] and on both being free from governmental influence (Interview D2, Oct 27, 2005). Members "shall not be bound to external instructions. They shall represent their personal convictions and shall be subject to their conscience alone".[125] 'Independence' also meant that members were "not to be associated with a specific position" (author's translation, Interview D2, Oct 27, 2005). The chair of the council stated that it was not the task of the council to provide "definite answers" but that it was "an authority for inciting discussion" (author's translation,

120 Several newspapers commented on the NER being established, such as: Die Welt, May 18, 2001; Kölner Stadtanzeiger, May 17, 2001; Berliner Morgenpost, May 15, 2001; Frankfurter Allgemeine Zeitung, May 15, 2001.

121 German National Ethics Council, Draft Decree submitted to Cabinet on 25 April 2001, Establishment of a National Ethics Council, at: http://www.ethikrat.de/_english/about_us/decree.html, Feb 23, 2006.

122 Ibid.

123 Ibid.

124 Ibid.

125 German National Ethics Council, Rules of Procedure, at: http://www.ethikrat.org/_english/about_us/rules_of_procedure.html, Feb 23, 2006.

ibid.). As a result of intense conflicts surrounding setting up the board and problematisations of the 'old discourse' as 'too emotional' or 'irrational', the council explicitly distanced itself from adversarial, political conflicts, from actors taking a definitive stand, and from taking a stance that was understood to be commonly binding. In the view of the NER, the main prerequisite for a "political solution" to the controversy was a

> "culture of mutual respect, in the spirit of which due regard is paid to divergent opinions and all arguments are objectively examined. Each side must be given an opportunity to seriously defend its position. Another element of mutual respect is the avoidance, in the course of discussion, of language likely to hurt, demean, or compromise the other" (German National Ethics Council 2001, 11).

Consequently, in its position paper on embryonic stem cell imports and research (ibid.), which will be discussed below, the council presented different 'assessment options' for discussion rather than recommendations for policy-making.

Summary: Problematisations of Discourse Practice

All three institutions/processes were the result of or an answer to problematisations of public discourse concerning bioscientific developments. While Fischer explicitly saw the symposium as a solution for the 'minefield of debate' on bioethics, she likewise clearly recognised the limits of discourse for finding *common rules*. From her point of view, there was not only a need to come to a decision, but, additionally, this task was in the hands of the political decision-makers. Fischer considered discourse to be (at least partly) a *prerequisite* for a political decision, but it was not meant to replace it. Nevertheless, the symposium was, in her eyes, an attempt to tame the antagonistic character of the debate and in this respect was not simply a prerequisite for politics but was itself a tool in conflict 'solution'. Fischer believed in a 'neutral space for discourse' in which the conflicting elements of the debate could be overcome by relying on the 'forceless force of the better argument' (Habermas 1995, 144). This 'solution' meant that conflicting elements and power relations and interests were dislocated and could not be discussed. Fischer's initiative, however, was stopped in January 2001. At this time she not only resigned, but also government actors engaged in the debate, leading to the transformation of problematisations and, furthermore, stimulating a *new* discourse, which included having the

NER established. The new discourse was both intended to replace the 'old' discourse and to replace a decision.

The parliamentary Enquête Commission was the result of deep conflicts on issues relating to biomedical developments, and it was established to meet the demands of the techno-sceptical anti-bioethics alliance that came into being in the 1980s. Against the background of problematising the lack of democratic legitimacy because of a lack of public discourse, the alliance explicitly demanded an institutionalised parliamentary discourse as a cure. The techno-sceptical movement hoped that the institutionalised discourse would support their demands (especially the demand to not sign the European Bioethics Convention), a view that was obviously shared by techno-optimists who consequently opposed its establishment. Therefore, establishing the EK REM initiated problematisation of the discursive dimension of the conflict. It provoked a debate on the function of discourse (e.g. the question of whether or not the EK REM would unnecessarily postpone a decision and the demand to stop discourse), the questions of who should participate in the discourse (e.g. the accusation that 'science was underrepresented'), and the question of how different actors might direct discourse in a certain (undesirable) way. The EK REM, it can be concluded, was understood to be a 'struggle commission' in which different actors hoped to be able to promote their views on biomedical issues. The assumed techno-sceptical impact of the enquête was taken seriously and a *new* discourse, a *new* institution, and *new* participants were demanded by techno-optimist actors.

When the new health minister took up her office and Chancellor Schröder engaged in the debate, problematisation of discourse intensified and was accompanied by calls for *new* discourse. Affirmed by government authority, a powerful counter-discourse to the kind of public discourse that took place the years before was initiated, mainly by techno-optimist actors. Nevertheless, Schröder had to face strong critique of what was interpreted to be authoritative intervention in the discourse, which should have been the task of parliament. As a matter of fact, setting up the NER was intended to initiate a 'more rational' and 'less emotional' debate than the 'old' debate. As Ulla Schmidt explained, it was to *replace* (restrictive) decisions on ES cell research and was to replace the implementation of a new act on reproductive medicine. All in all, the new discourse, which was initiated primarily by members of the government and accompanied by setting up a new institution, can be interpreted as a case of discourse stimulation by powerful social actors. The board did not simply have to prepare legislative action, as a permanent body its main task was to foster dialogue and public debate. In itself it was an institution for the stimulation of discourse. The NER was not only to advise the government but also—and more importantly—initiate, in Schröder's words, a 'truly public debate' as opposed to a debate in 'closed circles'. In this respect, the NER did not just have the task of organising and fostering public debate. It likewise provided 'information material', which included 'scientific details' or information on 'therapeutic and economic benefits'. A result of the conflicts related to the discourse, which comprised questions of who should participate or who can (legitimately) intervene in the debate, were declarations of the NER's 'independence'. It was stated that it was independent of the government and could not be associated with definite standpoints. The council explicitly distanced itself from directing decision-makers or the public on how to deal with the issues at hand or from committing itself to a (commonly binding) position on science policy. Instead, the discourse on the issues and the stimulation and organisation of discourse became the council's most important task.

A target of all three instances of institutionalised discourse was to structure or reorganise (public) debate. They were in themselves attempts to resolve the deep conflicts surrounding new biomedical practices by means of reorganising discursive practice, whereby they were initiated on the basis of various assumptions: the symposium was led by the hope of being able to provide neutral ground for discussion, free from adversarial elements; the EK REM was a struggle commission, which actors (especially techno-sceptical actors) thought would strengthen their position. However, the NER was established principally for the sake of discourse—for its own as well as for the stimulation of public discourse—in which it was not possible to proffer a definite solution or fight for a particular standpoint. On the contrary, it was assumed that each and every position must be respected. As will be shown in the next chapter, such an understanding of public discourse formed the basis for the statements on embryonic stem cell research of both boards, the NER and the EK REM. Problematisation their statements dealt with was guided by the assumption that, in principle, irreconcilable but equally valid viewpoints (on the status of the embryo) existed. This necessitated debate, although they could not be reconciled with one another.

4.8 Advisory Statements: The Status of the Embryo and Freedom of Research

In November and December 2001, the National Ethics Council (German National Ethics Council 2001) and the Enquête Commission (Deutscher Bundestag 2001) published their statements on imports of human embryonic stem cells. Both commissions had very little time in which to provide statements on the issue as soon as possible, that is, before researchers or the DFG simply created facts by importing ES cells. Despite the different processes leading to the founding of the two commissions and divergent members and institutional affiliations, they displayed several distinct similarities to one another, such as their respective definition of the origins and interpretation of the problem. Problematisation of a conflict between the status of the embryo and the freedom of research had prevailed in the debates of the last seven years. The following is an analysis of the statements by the two bodies with special emphasis on convergences within the two reports.

Science or Ethics?

Both reports were based on the assumption that the need for 'ethical' evaluation of stem cell research and imports arose from developments and prospects of science and technology and their potential applications.[126] "The ethical issues", the NER wrote, "are presented by the development of the life sciences" (German National Ethics Council 2001, 7), and the EK REM began its report with a detailed presentation of techno-scientific 'facts' of embryonic stem cell derivation, cloning, cell nuclear replacement, so-called risks and problems, as well as options for applications of these techniques (Deutscher Bundestag 2001, 8-20). Besides the current development, "the expected *future* developments" in the research (author's translation, ibid., 16, emphasis added) or the 'hoped for' results of experiments that may "one day" be arrived at (German National Ethics Council 2001, 9) formed the background of the 'ethical' evaluation in both reports. Directed explicitly at future developments, the then chair of the NER asserted that the council was to concentrate on "prognoses" and only to a

126 Parliament had already emphasised that, as a result of 'developments in the field of genetics', there was a need for 'ethical evaluation' of embryonic stem cell research or, more generally, the life sciences (BT DR 14/6551).

lesser degree on "day-to-day business" (author's translation, Interview D2, Oct 27, 2005). The inclusion of uncertain—i.e. "expected" or "hoped for"—*future* developments indicates that, from the outset, both commissions excluded the restriction of science and that it might *not* progress. Likewise, the presentation of 'scientific facts', especially in the EK REM report, suggested that, in principle, science was seen as providing normatively neutral and objective 'facts' for an 'ethical' discussion. The (not quite) hidden agenda was inevitably 'progressing' and 'sound' science. In this way, a more general inquiry of the direction, priorities, or expectations of scientific endeavour—i.e. the "implicit ethics" (Carr/Levidow 1997, 29) of science—was, in principle, excluded from the start. In addition, the underlying assumption of the neutrality of science accounts for the fact that the commissions' statements referred to the scientific differentiation between totipotent and pluripotent embryonic cells in order to distinguish between 'ethically doubtless' and 'ethically problematic' forms of research (cf. Deutscher Bundestag 2001, 51; German National Ethics Council 2001, 9). As a result, both commissions did not see any fundamental 'ethical' problems in embryonic stem cell research itself (if performed on *pluripotent* cells). Instead, they maintained that the 'ethical' problem arose from the origin of the cell, that is, from the embryo (Deutscher Bundestag 2001, 51; German National Ethics Council 2001, 9). Admittedly, the 'ethical' problem that was presented by the commissions also had a *legal* dimension: The Embryo Protection Act does not allow the derivation of ES cells, whereas neither the import nor use of these (pluripotent) cells was regulated and nevertheless considered (by some) to be allowed. For the EK REM, the 'ethical problem' had its source in the "moral intuition" of unease and rejection in segments of the German public, who regarded ES cell imports being permitted not only as "legally contradictory but as ethically doubtful double moral standards" (author's translation, Deutscher Bundestag 2001, 51). As a solution to this legal 'inconsistency', the NER stated: "The decision on whether to impose a legal prohibition or veto based on research-policy considerations, or expressly to allow imports, is primarily an ethical issue" (German National Ethics Council 2001, 10).

The 'Real' Problem: The Status of the Embryo and a Plurality of Views

While (current and future) techno-scientific developments were interpreted as presenting 'ethical issues', they were not the object of evaluation in the respective commissions' reports. Instead, both commissions presented the issue of ES cell imports and research as *ethical problems* that were related to the *moral status of the embryo*. Accordingly, both commissions assumed that public conflicts on the issue were the result of disagreements over this status. In the words of the EK REM, the "ethical evaluation of research on stem cells derived from embryonic tissue is impossible without recurring to the various views regarding the moral status of the embryo, especially in its early developmental phases, in order to examine which ethical convictions are relevant for law and legal development" (author's translation, Deutscher Bundestag 2001, 30). For the NER, the question of the use of embryos in research was "a matter of political and social controversy", deriving from the fact that "there is disagreement on the extent of the protection to which human life is entitled during its early embryonic development" (German National Ethics Council 2001, 11).[127]

Thus, while techno-scientific development was interpreted as both triggering the debate and as a resource for defining the object of the debate, both commissions problematised embryonic stem cell imports and research in Germany as a *conflict deriving from the plurality of moral views concerning the status of the embryo*. Consequently, the reports discussed the 'status of the embryo' at length and presented possible or existing perspectives on this status and staged debates on the issue within their statements.

The EK REM devoted several pages to the presentation and discussion of views on the status of the embryo, which, in the words of the commission, "exist in our society" (cf. Deutscher Bundestag 2001, 30-34). The commission distinguished between two general positions: The EK REM defined the first position as acknowledging the human dignity of the embryo and an end in itself from the time of fertilisation onwards, and not subject to balancing processes. The second position maintained that the human embryo has a graduated right to protection whereby the full right to protection of life applies only at a specific stage of embryonic development. A third position focused on the social relations in which the embryo is embedded, the feminist or ethics of personal relations (*Beziehungsethik*) position (ibid., 32f).

127 In another instance the NER declared: "The first relevant aspect is that of the permissibility, as a matter of principle, of using embryos for research projects in which these embryo are destroyed in the first few days of development. For this reason the National Ethics Council includes these considerations of principle in its deliberations" (German National Ethics Council 2001, 10).

In accordance with problematisation of embryonic stem cell research as an 'ethical' and not a political problem, the EK REM refrained from naming representatives of the above-mentioned positions, despite the fact that the issue was obviously a virulent political conflict in Germany. Framing the issue as a conflict that derives from 'different moral views on the status of the embryo' led to the assumption that the conflicts are irreconcilable (ibid., 34). In a further chapter, the EK REM provides a constitutional discussion on the 'status of the embryo' because the various 'ethical' positions were unable to reach a consensus that could be used as a basis for legal regulation (ibid., 34). The basic rights that apply in the case of embryonic stem cell imports, according to the commission, are the right to freedom of research (Art. 5(3/1), GG), the protection of human dignity (Art. 1(1), GG), and the right to life and bodily integrity (Art. 2(2/1), GG).[128] The commission stated that research on imported cells, provided they are pluripotent, is covered in the constitution by the right to freedom of research, which applies without reservation. However, other basic rights (especially the right to protection of life and human dignity) could restrict the right to freedom of research (Deutscher Bundestag 2001, 53). Restrictions could apply, for example, in a situation where two (or more) basic rights conflict with each other. While the EK REM saw no problem in importing pluripotent cells, the problems posed by imports arose from the fear that this would lead to the acceptance of the *derivation* of ES cells (ibid., 53). In light of all this the commission found it difficult to decide if a general prohibition of imports could be legally implemented, and the EK REM, in reference to the constitutional scope of regulatory interventions, failed to find a solution (ibid.). Whereas the constitutional issues lay in the hands of the legislator, "the Enquête Commission focused on deliberating the ethical questions that are vital for the legislator's evaluation" (author's translation, ibid., 54). However, the EK REM stated that 'ethical' questions cannot be resolved as "it is unforeseeable when dissent over the moral status of the human embryo in vitro will be eliminated in the future" (author's translation, ibid., 43). Thus, while 'ethical' questions were declared to be crucial—also in the case of legal regulation—they were presented as irreconcilable. In fact, the possibility (at least in the foreseeable future) of arriving at a mutual solution to the 'ethical' question of the status of the embryo was excluded.

---

128 Basic Law for the Federal Republic of Germany, at: http://www.iuscomp.org/gla/statutes/GG.html, Apr 3, 2006.

The NER report presented the results of discussions held within the commission itself. The first chapters provided statements and arguments shared by the members of the NER,[129] and in the following chapter the report presented two distinct positions on and discussions of the 'moral status of the (early) embryo', which members had disagreed on (German National Ethics Council 2001, 14ff). The NER presented a pro-research position that centred on an absolute understanding of the constitutional right to freedom of research and an alleged[130] 'right to health' versus the status of the embryo. It likewise presented a contra-position that was based on a defence of an absolute right to life of the embryo, which overrode any other constitutional rights including the freedom of research (ibid., 14-39). Consequently, both positions on the derivation of ES cells presented by the NER came to incompatible results. The main arguments of the pro-research position were that the German legal order was already based on a graduated right to life before birth (*abortion trap frame*) and that it must be possible to balance the protection of life versus other "obligations and objects of protection", that is, versus freedom of research and the "promising medical prospects" (ibid., 21) of embryo research. The pro-position within the NER concluded that the "freedom of research forbids the National Ethics Council—as it would a parliamentary commission of inquiry or the legislator—to dictate research workers the subject of their investigation" (ibid., 26f). To put it bluntly, science and research were presented as issues outside of societal or political intervention and, indeed, as issues that can only be legitimately regulated by science itself.

The second position presented within the NER against the derivation of ES cells was based on the assumption that the embryo has the absolute right to life. According to this position, 'life' begins with fertilisation and does not develop gradually *into* a human being but develops *as* a human being (ibid., 28). It is maintained therein that protecting 'human life' is the fundamental objective of the German legal order and that it is the prerequisite for realising any other objectives of human beings in accordance with their personal hierarchy of values (ibid., 30). The second position, in

129 These chapters described the function of the NER, the subject of the opinion, the normative framework, and the need for a culture of mutual respect.

130 The claim of a 'right to health' in the German constitution recurred throughout the debates on embryonic stem cell research in Germany, especially in arguments by techno-optimists. It was based largely on an interpretation of the article on the right to personal integrity or the right to life. Other actors, however, argued that the constitutional right to bodily integrity did not entail a 'right to health'.

turn, concluded that the freedom of research finds its legal limits, although these are not set by ordinary law but by other constitutional rights, and especially the right to life. The fundamental ethical principle, which holds the sacrifice of human individuals for the benefit of others as unethical, could not be ignored without jeopardizing the German legal order (ibid., 39).

These two statements on the status of the embryo formed the background for the presentation of possible 'arguments' concerning embryonic stem cell imports: The NER did not *discuss* the issue of imports but, instead, provided 'text modules', which readers could and should combine according to their individual viewpoints. The NER presented several 'text blocks' (subdivided into arguments pro or contra imports) "explicitly *without* evaluation" (ibid., 40, emphasis added) so they could be variably combined. Readers could choose from pro-import arguments, such as 'embryonic stem cells are not embryos and do not have to be protected for their own sake; a prohibition of import would mean interfering with the right to freedom of research'; 'a ban on imports would mean that German research could not contribute to the development of knowledge in a field that might be of great medical importance' (ibid., 41-43). On the other hand, the contra-arguments included the following examples: 'imports have to be prohibited as the underlying action—killing embryos—is ethically unjustifiable'; or 'imports imply that the production of embryonic stem cells in Germany meets with moral approval ' (ibid., 43-45).

The explicit conception of these argumentation 'chunks' as being "without evaluation" was not only based on problematisation of ES cell research and imports as a conflict over irreconcilable ethical perspectives on the status of the embryo, but also underpinned this problematisation. The NER expressly refrained from evaluating arguments in terms of right or wrong. In fact, by framing conflicts as *irreconcilable*, it excluded such judgments altogether. This quite clearly refers to the function ascribed to the NER more generally: to trigger and organise public debates while avoiding normative commitments. *All* arguments, positions, or beliefs in (public) 'ethical' debate 'must be respected' and 'objectively' examined (ibid., 11). 'All' is restricted to those arguments abstaining from assertions of being 'right' or 'wrong', and therefore no arguments were included that refused to 'respect any other argument'. The NER maintained that a decision on whether or not to allow imports "is primarily an ethical issue" (ibid., 10) "need[ing] to be settled by public debate" (ibid., 7) and this 'settlement' did not include a definite answer.

A Solution to the 'Ethical' Conflict: Public Debate

Despite assertions that it was impossible to find a consensus on the 'moral status of the embryo', both commissions nevertheless had the task of advising decision-makers. Consequently, both reports conclude with recommendations. As could have been expected, however, they did make any consensual statements and, instead, presented several "assessment options" (German National Ethics Council 2001, 46) or votes (Deutscher Bundestag 2001, 57ff). These ranged from calling for not only imports to be allowed but also for derivation of ES cells, allowing imports for a certain time and under certain circumstances (such as informed consent of 'donors' or a deadline for derivation), demands for a moratorium on imports until the legislator reached a decision, or demands for the prohibition of the use of embryos or embryonic cells in research (ibid., 57ff; German National Ethics Council 2001, 46ff). Altogether, the commissions acted out what they had suggested for (public) debate on 'ethical issues' more generally: They demonstrated that the respective 'moral' views were irreconcilable and that it was impossible to find a common/definite solution to the 'ethical problem'. Either despite or because of the assumption of a definite answer to the 'ethical' conflict being impossible, both commissions considered a certain 'solution' to be appropriate. This was a *practical* 'solution' and not a *substantial* one: public debate.

As had been said earlier, both commissions were set up in a situation in which public debate was increasingly problematised and a new discourse (in the form of the NER) was demanded. In confronting this situation, their explicit target was to organise, structure, and deepen, that is, to stimulate public discourse. As the decree on establishing the German National Ethics Council stated, it "shall organize the social and political debate and ensure that all the relevant groups are involved. It shall provide citizens with information and material for discussion (e.g. exhibitions, publications, internet forums etc.)",[131] and the EK REM will prepare statements for the "ethical evaluation" of "medical future questions" in order

131 German National Ethics Council, Draft Decree submitted to Cabinet on 25 April 2001, Establishment of a National Ethics Council, at: http://www.ethikrat.de/_english/about_us/decree.html, Feb 23, 2006.

to "deepen public discourse and to prepare political decisions" (author's translation, BT DR 14/3011).

Recommendations or statements by the commissions were neither confined to giving advice to policy-makers nor predominantly were made for this purpose. Rather, they were specifically meant to stimulate public debate. On the surface, this focus on extending the debate to the public sphere suggests that the model of "best expertise" (Irwin 2006, 303) in science policy was losing ground. It intimates a departure from the conjecture that expertise, science, and scientific knowledge *alone* can supply appropriate solutions to the problems at hand and presents public debate as the answer. In fact, we find in both commissions what Irwin and Michael (2003) and others (cf. for example, Carr/Levidow 1997; Irwin 2001; 2006; Wynne 2001) detected in science policy more generally: a trend towards more inclusiveness, an emphasis on the need to integrate the public, and fostering dialogue between science and the public. While this superficially implies a departure from the belief in the supremacy of scientific epistemology, public debate does not have the goal of replacing scientific expertise and expert advice but of accompanying it. Both statements quite clearly advocate both a considerable confidence in the 'neutrality' and 'objectivity' of science or a 'facts first' approach and the need for public debate. As the NER wrote:

> "The function of the National Ethics Council is to express views on the ethical issues that are presented by the *development of the life sciences* and need to be settled by *public debate.* In doing so it must assess *the possible consequences of scientific and technological innovations,* evaluate them in accordance with *universally evident criteria,* and reflect on the significance of *the chances induced by science and technology* both for the living conditions of the individual and for society at large" (German National Ethics Council 2001, 10, emphasis added)

The EK REM stated:

> "The importance of these ethical questions reveal how decisions on legal regulation can only be made if the aforementioned *natural-scientific and medical conditions* have been sufficiently clarified and only after *intensive public debate*" (author's translation, Deutscher Bundestag 2001, 43, emphasis added).

Public debate was (and, in fact, still is) presented as a necessary supplement to scientific evidence albeit clearly separated from it. Additionally, it is portrayed as something distinct from science: The obvious result is that, in public debate, questioning 'sound science' is severely restricted because it is not politicised. Also the fact that both statements were based on the general assumption of scientific 'progress' being inevitable supports such an interpretation. At the same time, public debate was not depicted as a way of finding a mutual solution. Both commissions recognised that it was neither possible nor necessary to reconcile the controversy in their own deliberations or in public debate as far as problematisation of a plurality of *irreconcilable* 'ethical' views on the status of the embryo was concerned. Public debate is simply a space in which *ethical* controversies can be expressed but not solved. But, if public debate is neither meant to challenge the model of 'best expertise' or scientific development nor to reconcile 'moral controversy', what kind of 'solution' to 'which' problem is public debate? In this regard the NER was most explicit:

"*Everyone must* be able to form an impression of the prospects and risks of the new technologies, as a basis for arriving at *his or her own judgement* on the associated ethical issues. To this end, the Ethics Council will seek to facilitate understanding of the presuppositions and consequences of current problems" (German National Ethics Council 2001, 7, emphasis added).

The NER considered the stimulation of public debate to be an invitation to everyone to deal with and think about 'ethical issues' arising from the possibilities of new technologies and to form their own opinions on them (cf. also Herrmann/Könninger 2008). This very liberal and individualistic understanding of 'ethics' is based on and supports the refusal to give 'definite', mutual solutions in science policy, and upholds the conjecture that moral controversy is 'irreconcilable'. The fact that everyone "must" be able to participate at the same time suggests that the public has a "responsibility to participate" (Kerr 2003, 217) and engage in 'informed' debate, and not that it can or has the right to do so. The public must not, however, have the prospect of challenging 'scientific progress' or of providing definite solutions to the 'moral' controversy. Indeed, "the prospects and risks of the new technologies" are the preconditions for public debate.

Summary: Ethics, Science, and Public Debate

The German National Ethics Council and the Enquête Commission on Law and Ethics of Modern Medicine had been established at a time when public debate on imports of and research on human embryonic stem cells was at its peak. At the same time, the plurality of problematisations had seriously decreased: Public debates now focused on problematisations of a conflict between the 'moral status of the embryo', its human dignity, and right to life as opposed to the freedom of research. Although the commissions stressed that 'ethical' issues would arise from techno-scientific developments (now and in the future), these were not the object of their deliberations. Instead, they problematised the 'moral status of the embryo', to be more precise, a *plurality of moral views* concerning this status, and consequently they presented in their statements so-called "main positions" (German National Ethics Council 2001, 14) or perspectives "existing in our society" (Deutscher Bundestag 2001, 30) on this status. However, they refrained from naming the representatives of such positions, as had been the case also in the British CMO report (cf. 3.7), whereby they disclaimed a yearlong, very controversial public debate on biomedical developments and did not acknowledge the adversarial character of these debates or the conflicts concerning problematisations. In this way, the commission fell in with the rejection and disqualification of the 'old' debate and calls for a 'new' debate, which came up early in 2001. This process was accompanied by the semantic narrowing down of problematisations. As a result, both commissions had the task of stimulating and organising public debate, a demand that implied problematisation of the 'old' debate. They additionally expressly recommended more public debate. Despite both commissions differing as regards the conflicts surrounding their establishment or their institutional affiliation, one being a parliamentary 'struggle commission' connected to the techno-sceptical alliance, the other being the Chancellor's 'forum for dialogue', they adopted and institutionalised the semantic as well as the practical dimension of the dominant discourse.

The discourse was based on separating science, facts, and expertise, on the one hand, and the public, ethics, and moral views, on the other. At the level of problematisation, the commissions clearly differentiated 'ethical' issues from scientific evidence and neutral and objective facts. Conflicts, in their view, arose from a plurality of irreconcilable *ethical* views concerning the status of the embryo, whereas they presented scientific 'progress' and the 'neutrality' of science as acknowledged facts. By differentiating in this way, the 'objective' and supposedly 'neutral' scientific dimension was not reflected on (Carr/Levidow 1997; Wynne 2001) and scientific development appeared as the indisputable *prerequisite* for public 'ethical' debate—a prerequisite that was outside of the scope of problematisation and political intervention. Additionally, by separating 'science' from 'ethics' and problematising irreconcilable ethical views on the status of the embryo, a new and endless space for discussion was established: Problematisation of *irrec-*

oncilable ethical views not only supported calls for more public debate but precluded the prospect of ever arriving at a common or definite—or even a less restrictive—answer. Precisely the fact of not being able to find a definite answer underpinned calls for 'an intensive public debate' as an answer to (but not for finding a resolution for) 'ethical' issues. More generally, public 'ethical' debate or 'ethical' discourses were to provide a space for the *articulation* of different perspectives (on the status of the embryo) without the prospect of ever arriving at a common solution. In practice, 'discursive resolution' is a vicious circle of never-ending 'ethical' debate accompanying never-ending scientific 'progress'.

4.9 Decision-Making Case by Case

In the run-up to the parliamentary decision due at the end of January 2002, four parliamentary motions existed: One that demanded a total ban on embryonic stem cell research in Germany, including a ban on imports; a second one that promoted a general prohibition of imports but allowed certain regulated exceptions; a third motion supported the approval of imports, given that 'surplus embryos' were used; and a fourth motion promoted a general approval of stem cell research, including production and imports of ES cells and, therefore, a change of the EschG. By the time of the final debate, motion three and four were fused into one that incorporated a general approval of imports and was open towards the production of ES cells also in Germany without categorically including it.

On 30th January 2002, the German *Bundestag* decided in a free vote between three motions: (1) an absolute ban on the import of embryonic stem cells (BT DR 14/8101); (2) a ban on embryo research, prohibition of imports of embryonic stem cells whilst allowing restricted exceptions (BT DR 14/8102), [132] and (3) permitting research on embryonic stem cells and permitting import (BT DR 14/8103). The third proposition was outvoted in the first ballot. In the second ballot, MPs voted for choice No 2: the prohibition of embryo research and the prohibition of the importation of ES cells. However, imports would be allowed in certain restricted cases, that is, if research projects meet several requirements: imported stem cell lines have to be established before a certain deadline, that is, before the vote in

132 In terms of contents, this motion very much resembled the minority vote of the EK REM (cf. Deutscher Bundestag 2001, 57ff).

the *Bundestag* (30th Jan. 2002) so as to ensure that imports do not cause the destruction of more embryos. Secondly, the embryo must be created to establish a pregnancy and not implanted for other reasons. Thirdly, donors have to agree to the embryo being used for the production of ES cells and must have no financial interest in the matter. Fourthly, the 'noble goals of the research' must be confirmed, and an ethics committee must monitor the ethical acceptability of the research project. Last but not least, a transparent and legitimate board was to control imports of and research on human embryonic stem cells. The decision to prohibit the import of ES cells while allowing exceptions on a case-by-case basis was widely understood to be a means of avoiding procedural bars: quite clearly the two more extreme propositions of the three motions faced difficulties in winning majorities (Interview D 8, Jan 19, 2006; Interview D 5, Nov 30, 2005). Many different actors—those opposing and those supporting imports—agreed to a 'no-but' decision, even if they basically only agreed in order to prevent one of the extremes.

A few days after the vote, the deadline for the establishment of ES cells that could be imported to Germany came under discussion: researchers from Bonn declared that the deadline meant not only 'quantitative' but also 'qualitative limitations' of ES cell research (Frankfurter Rundschau, Feb 4, 2002). It would impose a handicap on German researchers in international biotechnology, as researchers from other countries could work with 'newer stem cell lines' (ibid.). The president of the German Research Foundation added that he wanted stem cell lines to be produced in Germany. Pro-import parliamentarians suggested the date of the implementation of the new act as a deadline (some time before October 2002), implying that from now on stem cell lines could be produced for the German market. Restricting research to existing ES cells was not be "ethically justifiable" as it would "hinder the research for treatments and therapies" (author's translation, ibid.). Parliamentarians who had voted for the clear restriction of import replied that they would never agree to a later date and that "researchers should not think they can break one barrier after the other" (author's translation, ibid.). The proposed bill was presented 22 February and suggests 1 January 2002 as a deadline.

After a revision of the bill on imports of ES cells, the Stem Cell Act (*Stammzellgesetz*; StZG 2002) was passed in March 2002 and came into effect 1 July 2002. For the time being, an end was put to the public debate on embryonic stem cell research and imports when the law was introduced and the Central Ethics Commission for Stem Cell Research (*Zentrale Ethik-Kommission für Stammzellforschung*, ZES) was established.[133] The ZES is an expert body, which includes experts from the life sciences, ethics, and theology. It is the first *ethics* body in Germany that actually has the authority to make decisions.

Today, the regulation of embryonic stem cell imports in Germany follows a mode of rule and exception, which in effect is case-by-case decision-making resembling the HFE Act. Rather than merely prescribing what is and what is not allowed, the Stem Cell Act (StZG) and the related statutory order[134] determine the arguments, discourses, and activities that must be produced or presented by applicants, by the ZES itself, or by intermediate bodies in order to obtain approval for the proposed research projects. The ZES then decides which activity is considered 'ethical', and can hence be approved, and which is not. A positive statement by the ZES is a prerequisite for approval (StZG 2002, Art. 6. Abs.4).[135] Therefore, the goal of the law is not to prohibit or allow certain activities in general (that is, the import of and research on embryonic stem cells), but rather to locate them in a system of rational argumentation, deliberation, and management. Decisions are put into the hands of an expert body—the ZES—that is, of a body of professional self-regulation rather than state intervention, although 'professional' here includes also 'ethics experts' and theologians.

Thus, the German Parliament, in its attempt to 'resolve' the conflicts, neither prohibited nor generally allowed stem cell imports and research. Parliament decided, on the contrary, to refer decisions on imports to an expert body distant from the state. The roles of parliament and of the law in science policy-making were decisively transformed into *framing* the terms that must be used and the activities that must be performed by the actors involved. Actual decision-making was the responsibility of the ZES. To put it succinctly, parliament provided the rules and frames for the discourse and the ZES administered and controlled adherence to these rules. As was the case in Britain, in this respect framing the issue as an 'ethical'

133 The webpage of the ZES can be found at: http://www.rki.de/cln_048/nn_207098/DE/Content/Gesund/Stammzellen/ZES/zes__node.html?__nnn=true, Oct 19, 2006.

134 Zentrale Ethik Kommission zur Stammzellforschung, Statutory Instrument, at: http://www.bgblportal.de/BGBL/bgbl1f/bgbl102s2663.pdf, Oct 19, 2006.

135 StZG (2002). *Act Ensuring Protection of Embryos in Connection with the Importation and Utilization of Embryonic Stem Cells (The Stem Cell Act) of 28 June 2002 (unofficial translation)*, http://www.bundestag.de/parlament/gremien/kommissionen/archiv15/ethik_med/archiv/stammzellgesetz_engl.pdf , Feb 8, 2007.

one was consequential: Analogous to understanding the function of public debate as a discursive space in which to *articulate* different moral perspectives without the prospect of finding a definite resolution, parliament reclaimed state neutrality towards 'morals' and refrained from giving an absolute answer to the problem. It withheld making a clear decision on whether imports were allowed or not and, instead, implemented a deliberation process on the 'ethical quality' of the respective research projects and implemented a decision-making process on a case-by-case basis. Framing the issue in terms of 'ethics' and by referring it to an 'ethics body' signalled that 'ethical concerns' were taken seriously and measures of 'control' were implemented, simultaneously ensuring that scientific endeavour was able to proceed. Indeed, in April 2008, the deadline for the production of embryonic stem cell lines that Germany can legally import was postponed to 1 May 2007.

4.10 Conclusion: Germany—A Struggle over Problematisations

In Germany debates on reprogenetics took place against the background of a rather strong anti-biomedicine and bioethics alliance that was formed at the end of the 1980s. This movement not only problematised biomedical practices but also a certain type of discourse on the same. Therein, utilitarian bioethics discourse was rejected and a counter-discourse based on human rights and dignity promoted. Such initiatives were pivotal for the publication of the hitherto arcane activities of policy-makers (i.e. the drafting of the European Bioethics Convention) and for initiating and calling for public debate on biomedical issues. In the mid-1990s, however, a new pro-research debate was initiated by another group of actors, namely professional organisations and medical professionals endorsing the deregulation of embryo research and demanding a cutback of state rule in favour of professional rule. These initiatives provoked strong reactions in the (techno-sceptical) public and can be interpreted as instances of discourse stimulation. Techno-sceptics tried to counter discourse and problematisations introduced by professional actors by presenting their own interpretations of the problems involved in PGD.

The debate in Germany on new reprogenetic practices, such as pre-implantation genetic diagnosis or embryonic stem cell imports and research, was highly controversial from the very beginning. Opinions in this field were extremely polarized: policy-makers and the public were divided into opposing camps that were unable to reach a consensus. One group sought more liberal legislation for PGD in particular or human embryo research—practices, which in their view, could benefit many people. The other group stated that the new techniques were essentially attacks against fundamental moral principles, such as human dignity or non-discrimination. Indeed, we can speak of a "wicked problem" in Fischer's (2000, 128) terms: meaning that different actors provided quite different problematisations related to the issues at hand. The German situation can be described as a *struggle over problematisations*, a struggle over the frames, terms, and discourses on these issues: ranging from problematisation of social conditions and consequences of scientific research to problematisation of a conflict between restrictive law and personal or scientific interests, or between economic prospects and the status of the embryo.

When the discourse came into full force in 2000/2001 it was accompanied by several changes at the semantic as well as the practical level: especially the interventions of members of the government (Chancellor Schröder and Minister of Cultural Affairs Nida-Rümelin) led to the scope of problematisations being narrowed down, which brought the focus on the status of the embryo. Especially this focus and questioning the embryo's human dignity functioned as a catalyst for the debate. This proved to be an incentive for techno-sceptical actors to engage in the debate and, more importantly, to frame their arguments in terms of the embryo's status. Techno-sceptics now felt compelled to frame their arguments in these terms, whereas they had previously framed the problem in socio-political terms and now problematised the status of science. This problematisation, however, was based on the imperative of scientific progress, which also served as the background for interpretations of human dignity or the right to life. On the grounds of the imperative of scientific progress, human dignity or right to life were framed as cases of a German *Sondermoral*, implicitly or explicitly pleading for the deregulation of science. Additionally, at the time discourse practice was problematised—in this case, however, not by techno-sceptical actors as in the 1980s but by techno-optimists. Already the anti-bioethics alliance of the 1980s demanded more discourse as a response to the *lack* of public debate and *lack* of techno-sceptical viewpoints in closed circle debates, whereas in 2000 techno-optimist demands for more discourse were directed at the rationalisation of public debate. Especially in the 1980s and early 1990s, techno-sceptics maintained that more discourse was inseparably linked to demanding the restriction of scientific endeavour, while calls for more rational debate in 2000 were based on the assumption that a finding a definite answer to the issues at hand was impossible. This understanding of (the goals and function of) public debate was linked to problematisation of biomedical developments in 'ethical' terms, i.e. concerning the status of the embryo. The official institutions established in order to deal with the issues (the EK REM and the NER) also adopted this problematisation, interpreting the political conflict as an 'ethical' conflict between different irreconcilable views on the status of the embryo. In this way, the political conflict about the status of science in society was translated into an 'ethical' conflict over the status of the embryo, which prevented a more general questioning of scientific endeavour. Secondly, this problematisation underpinned calls for more public debate. At the same time, normative commitments had to be avoided, and consequently the prospect of ever arriving at a common definite answer to or seriously challenging scientific 'progress' was excluded. The final decision in parliament to refer decisions on research projects to an expert commission adheres to framing the issue as an 'ethical' issue concerning the status of the embryo and not an issue of political conflict over the status of science in society. Finding a 'resolution' to 'the controversy' in Germany was, in fact, done by providing a framework for the management of stem cell research applications in an 'ethical' manner, which means that applications and procedures of decision-making (by the ZES) must comply with certain discursive standards and parameters. In this way, ongoing scientific research was secured and at the same time calls for control and restriction were 'integrated'—even if only at a rather symbolic level—that is, by introducing a deliberation process that was led by an expert commission and by demanding and supporting an endless public discourse on 'ethical' issues of scientific developments. Such a discourse accompanies scientific 'progress' without challenging it.

Conclusion: Problematisation, Discourse Stimulation, and Ongoing Scientific Research

The analysis of public policy debates on new reprogenetic practices in Great Britain and Germany that took place between the mid-1990s and 2001/2002 provides significant insights into public engagement in science policy discourse and its effects on or function within science policy-making for a critical understanding of the same. Remarkably and contrary to demands or assumptions, in both countries public debate—and public 'ethical' debate in particular—did not form a counter-discourse to the imperative of scientific progress, but in fact strengthened it. In this respect, processes of problematisation played a decisive role: The ways in which reprogenetic science and practice was defined as problematical as well as how problematisations changed throughout the years influenced the direction of debate and the search for a 'solution'. Also, implicating the necessity to talk about and change the 'problem', problematisation must be understood as an act of discourse stimulation: Certain problematisations triggered certain discourses, inviting and directing participation in discourse. Two interrelated processes became apparent: Firstly, the semantic level of problematisation: how the issues were framed as policy problems, how problematisations differed between actors, and how (dominant) problematisations changed throughout the years. Secondly, how problematisations triggered certain discourses, who were the actors initiating the discourses, and with what effects. Both had particular power effects, which, in the two cases discussed, restricted participation in the shaping of science policy rather than supported it, while at the same time, stimulating public debate. Indeed, problematisation and discourse stimulation strengthened rather than challenged the commitment to inevitably ongoing scientific research.

Post-positivist, discourse-analytical, or interpretive approaches to public policy have shown that problematisations rest on interpretations of reality involving normative choices and judgements. They reject the existence of neutral and pre-set policy problems or the existence of a normatively neutral language to define these problems. Different problematisations imply various strategies in dealing with 'problems' and suggest different ways of solving them. Which problematisation gains dominance is therefore decisive, and problematisations and not solutions alone become objects of struggle. In studying policy-making in the two countries, we have seen that problematisations related to the issues of pre-implantation genetic diagnosis, embryonic stem cell research, cell nuclear replacement technology, or so-called 'therapeutic' cloning differed as far as which actors were concerned and as far as problematisations changed throughout the years. The new practices posed 'wicked problems' for policy-making, in which there was not only disagreement at the level of handling these issues but already at the level of problematisation.

In Great Britain problematisations ranged, for example, from 'public concerns about science potentially blocking scientific developments' to problematisation of an 'imbalance of public debate on the risks and benefits of the research', a 'conflict between the status of the embryo and the greater good for society', or problematisation of 'serious diseases'. In Germany, definitions could be found that ranged from problematisation of 'legal restrictions as research impediments' to 'a conflict between a German Sondermoral and the economic potential of embryonic stem cell research', problematisation of the 'negative social consequences of new research fields', or of 'the status of the embryo versus the freedom of research'—all of which implied certain normative assumptions and visions for finding 'solutions'.

If we consider the plurality of problematisations that appeared in both debates, we can speak of struggles over problematisations in both countries, albeit to a greater or lesser extent. In Germany, the struggle over problematisations was much more adversarial in character, with various actors promoting very different problematisations and concepts for finding a resolution. Germany's techno-sceptical community already attracted attention before the debates under study began. This community was formed in conjunction with the publication of the European Bioethics Convention in the early 1990s. Starting in 1995 with the publication of statements on embryo research and PGD by two professional scientific organisations, the German debate was marked by actors differing greatly on the question of what the problem 'really' was. In Germany there were two main opposing camps: One supported the liberalisation of embryo research and PGD and the other promoted legal prohibition. They respectively introduced diverging problematisations of the issues. Supporters tended to problematise legal restrictions of PGD or embryonic stem cell research as research impediments illegitimately constraining the freedom of research, as restricting the options and interests of concerned people, or as illegitimately inhibiting the economic potential of new research options. Opponents, on the other hand, problematised the social and political conditions and consequences of unconstrained research, the likelihood of slipping into increasingly dubious research, or the illegitimate instrumentalisation of women's needs in order to support scientific research. The plurality of problematisations, however, did not prevail. When the discourse came into full force in the years 2000/2001, we find the problematisations present in public discourse in Germany being significantly narrowed down. In 2000/2001, the dominant problematisation focused much more intensely on the 'status of the embryo', on the question of whether an embryo has human dignity and a right to to life. The interventions of the German chancellor and the minister of culture in particular triggered a narrowing down the scope of problematisations, which was accompanied by a massive increase in discourse. Problematisation of the 'status of the embryo' and rejection of its human dignity and right to life was an incitement for techno-sceptical actors to engage in this debate and to frame their arguments in these terms. Especially feminist techno-sceptics felt impelled to become involved in debates on the embryo, even if it was for (partly) strategical reasons. While discussion on the scope or legitimacy of the constitutional guarantee of the embryo's right to human dignity and to life gained dominance, other problematisations at the time faded into the background or disappeared altogether. A conflict was problematised between the 'status of the embryo and the freedom of research'—a problematisation that appeared as the problem as such and prevailed until the end of the debates. Established respectively in 2000 and 2001, two advisory boards, the German National Ethics Council and the Enquête Commission on Biomedicine dealt with this 'problem', framing societal conflicts as originating from various ethical viewpoints on the 'status of the embryo'. This focus also meant abstaining from problematisation of scientific development more generally. Techno-sceptical problematisations, such as those along the lines of the 'conditions and consequences of scientific endeavour', only appeared, if at all, sporadically.

The techno-sceptical community was evident in Germany and it constantly tried—though not successfully—to rebuff techno-optimist problem definitions and introduce problematisation of social conditions and consequences of science and research. This was, however, not the case in Great Britain. The 'struggle' over problematisations in Great Britain was rather a quixotic fight, highly exaggerating public concerns about or opposition to the new research. For example, in newspaper coverage of 'therapeutic' cloning or cell nuclear replacement technology, it was common to problematise the new research in terms of 'risks and benefits'. Moreover, there was a much greater number of supporting articles highlighting the 'benefits' of the research than critical ones. The 'struggle' over problematisations in Great Britain was rather a search for a problematisation to support the legalisation of 'therapeutic' cloning in face of anticipated or potential opposition. In the first consultation that took place a few weeks after the announcement of the birth of the cloned sheep Dolly, problematisation of 'public concerns' and of an 'imbalance of public debate' was presented before a debate had even started, let alone a controversial debate. 'Concerns' were highly exaggerated, and it was not quite clear who the dialogic partners, the representatives of these 'concerns', really were. Problematisation of 'concerns', however, remained stable in the debates, appearing in problematisations such as that of 'misinformed or irrational public debate' or of a 'deficit of public trust'. Underlying problematisation of 'concerns' was the assumption that scientific development was inevitable: The problem was, for example, interpreted as comprising public misunderstanding of the benefits and risks involved in the new research, as a public debate that overemphasised the risks and therefore needed to be re-balanced, or as a problem of uncertainty or misunderstanding within the public over regulating scientific endeavour. Problematisations of 'public concerns' in the UK imitated a debate whilst concealing a strong science bias. Therefore they can be interpreted as a way of signalling that a counter-balance to science was being integrated without this necessarily being the case. In the search for a problematisation to legitimate the legalisation of 'therapeutic' cloning, integrating a counter-balance to science apparently played a crucial role in signalling that the process of reaching this target was not dominated by science. In this respect, problematisation of 'serious diseases', as was presented by the Chief Medical Officer's expert group, was decisive: 'therapeutic' cloning—and the progress of science—were transformed into items that were of public interest and into the solution for the 'problem of serious diseases' that had no alternatives. Additionally, as was the case also in Germany, in the CMO report a problematisation of the 'status of the embryo' as *the* moral question gained dominance in interpretations of societal 'conflicts' or public concerns about 'therapeutic' cloning. The report, which also served as the background for the British parliamentary debates, presented the conflict on whether or not to allow 'therapeutic' cloning as a moral controversy on the status of the embryo. At the same time, it claimed that a discussion on whether or not embryo research should be allowed was beyond the scope of deliberations. As the analysis of the parliamentary debates has shown, the exclusion of a more general questioning of embryo research prevailed until the end and was a safeguard for supporters of the new research. Problematisation of 'serious diseases' formed the moral value that was to outweigh the status of the embryo, underpinning the necessity and inevitability of scientific development in the parliamentary debates as well: Both the supporters and opponents of research cloning argued in favour of more research, the former of embryonic stem cell research, the latter in favour of adult stem cell research. Thereby, discussing alternative 'solutions' to the 'problem' of 'serious diseases' was excluded and scientific developments were prevented from being problematised and politicised more generally.

In the policy debates under study we find similarities and differences in Great Britain and Germany on the issue of discourse stimulation. In 1995 in Germany, professional scientific organisations such as the German Research Foundation (DFG) or the German Society for Human Genetics (GfH) triggered a debate by problematising restrictive interventions of the state in science as 'research impediments'. Problematisation of state intervention was accompanied by a 'test case', i.e. the application of two medical professionals for permission to perform a PGD, a practice that many interpreted as prohibited at the time in Germany. The measures taken, such as triggering debate, problematising restrictions, and the test case, were related to the fact that, with the change of the constitution, wherein regulatory competences concerning health issues were transferred from the *Länder* to the *Bund*, the implementation of a new federal act on reproductive medicine became likely. This implied the possibility of either losing professional self-regulatory competences in the field of reproductive medicine or slackening embryo protection. Therefore, problematisation of 'unreasonable' or 'unethical' restrictions of research and practice were meant to stimulate a debate on the legitimacy of restrictive state intervention in science and its applications. To be more succinct, these problematisations were meant to trigger a debate on the Embryo Protection Act, which was directed at removing legal restrictions and strengthening professional self-control. They were also meant to set up a framework for discussion: The distinction between 'totipotent' and 'pluripotent' cells, framing PGD as a merely scientific matter, introducing the comparison of PGD and abortion, framing PGD as some kind of 'early abortion', or the balance between freedom of research and other values were crucial in this respect. Additionally, intervention on the part of the Bioethics Commission *Rheinland-Pfalz* in 1999 or the publication of guidelines on performing PGD by the German Medical Association in 2000 functioned as 'bases for discussion', stimulating and setting the terms for debate.

Especially in 2000/2001, Germany faced a massive increase in discourse on biomedical practices and research related to permitting imports and using human embryonic stem cells. These debates were again introduced by scientific professionals supported by members of the government and directed at achieving a cut-back of state restriction of science and research. Intervention on the part of Chancellor Gerhard Schröder and the Minister of Culture Nida-Rümelin that brought the focus on the status of the embryo was accompanied by discourse increasing enormously. Questioning the 'human dignity of the embryo' was a catalyst in the debates: it acted as an invitation or even demanded that one take a stand on its status and that one participate in this discourse. One result of the stimulation of discourse on the status of the embryo was that techno-sceptics were integrated in a discourse in which they had to accept the terms and problematisations presented by techno-optimists.

At about the same time, problematisation of discourse and calls for a new (more rational) discourse in Germany appeared. This new discourse was not only intended to replace the 'old' one but was to expressly serve as a substitute for arriving at a decision: setting up the German National Ethics Council and accompanying calls for more and more rational public debate had the target of stopping the legislative process that the former health minister had initiated. The council had the function of fostering dialogue and public debate and was itself an institution for the stimulation of discourse. Calls for new public discourse were of major importance in the statements of the NER and the EK REM. In combination with problematisation of the status of the embryo, such calls rested upon the assumption that moral views were principally irreconcilable and that it was impossible to ever achieve a definite solution. Nevertheless, both institutions suggested that more public debate was an answer to the 'ethical problem', a solution that, in practice, signified a never-ending 'ethical' debate accompanying never-ending scientific 'progress'.

In Great Britain, on the other hand, the debates on the HFE Act began with the announcement of the birth of the cloned sheep Dolly by the Roslin Institute and the immediate involvement of high-ranking policy bodies, such as the Science and Technology Committee, problematising the 'potential of public concerns to block scientific developments'. One of the crucial differences to the case in Germany was that already a few weeks after the Dolly was announced, high-ranking policy bodies were involved in and led the debates. Also, as has been said earlier, there was much less public debate in Great Britain and it was less controversial than in Germany. From the beginning in Great Britain, 'public debate' was problematised as being 'irrational' and ways were sought to 'rationalise', 'inform', and increase public debate on science. While the strength of public concerns about or critique of the new research were highly overemphasised by the respective bodies, problematisation of these concerns underpinned calls for more (and more rational) public debate. At the level of discourse practice, problematisation of 'public concerns' was a device for stimulating debate and activating public engagement, such as in the public consultation conducted by the HFE Authority and the Human Genetics Advisory Commission in 1997. A closer analysis of this consultation reveals, however, that public involvement had a great educational impetus and, in Alan Irwin's terms, served to set up a 'framework for engagement': For example, by excluding the 'old', more general questioning of embryo research and imposing a focus on 'therapeutic' cloning, the consultation required of participants that they must generally accept embryo research in order to be acknowledged as 'informed' participants. More general disagreements on embryo research were used to substantiate further calls for more public and institutionalised debate: General critique of embryo research was interpreted as beyond the scope of the consultation and a result of misinformedness. Also, the fact that the British government frequently rejected the reports provided by different advisory commissions in order to refer the issue to another advisory body can be interpreted as instances of stimulating institutionalised discourse, while calling for further debate justified the postponement of a political decision. With criticism of new scientific research options being rather minimal among the public, the government nevertheless claimed that it had to 'assure public confidence' in science and that further deliberation was needed so as to come to terms with mistrust. Therefore, the government took up a pro-active stand on 'therapeutic' cloning by stimulating an (institutionalized) discourse on the pros and cons of the new research in order to prevent conflicts before they actually arose. These discourses were not merely directed at finding a resolution to an existing conflict but at structuring the debate in order to predetermine problematisations against the background of the imperative of scientific progress, which was beyond question.

Towards the end of the two debates under study, we find that in both problematisation of a plurality of principally irreconcilable individual ethical perspectives on the status of the embryo gained dominance. In both countries this problematisation entailed not presenting real-life actors holding a certain perspective (for example, in the reports by the German National Ethics Council or the British Chief Medical Officer). This implied that 'ethical perspectives' were simply 'out there', independent of actors, interests, or political and social standpoints. Individual 'ethical' perspectives became the problem behind (existing or potential) conflicts. In this way, a potential or actual adversarial controversy on techno-scientific developments was framed as an issue concerning a variety of equally valid views from an unrelated aggregation of uninterested or 'innocent' individuals who needed to be integrated in the debate.

In Great Britain, this 'ethical' problematisation, focusing on individual moral perspectives, implied a refusal to principally restrict the research because limitations would imply a 'moral imposition' on the individual. Consequently, in Great Britain, the 'problem' was 'solved' by allowing the research while insinuating that the decision to use or support the research lay in the hands of the individual. Hence, science would have to pursue research further in order to provide a 'choice' of options, enabling individuals to choose whether or not to use research results. Regulation of 'therapeutic' cloning was framed as the result of personal moral decisions, while parliament abstained from finding a definite resolution. Instead, parliament was interpreted as being responsible for preparing the grounds for science to proceed so that it could provide options for individual decisions. The decision on research applications, however, was referred to the HFEA in order to signal that science would proceed under control. Likewise, in Germany the 'ethical' problematisation was linked to a refusal of committing oneself to a mutual normative answer. Here, the 'problem' was 'solved' by allowing imports and research, which was, however, accompanied by the establishment of a new 'ethics' body. This body had the task of deciding on applications for imports and research on a case-by-case basis, much like the HFEA. Also in Germany, parliament refrained from giving a definite go-ahead to research or explicitly forbidding it. Therefore, both parliaments abstained from giving a definite answer to the question of whether or not the new research is permitted and referred case-by-case decisions to a commission far removed from the state and to the individual consumer. By this move, both secured that scientific research could go on, in effect answering definitely for research. At the same time, both countries signalled that 'ethical concerns', 'opposition', or calls for restriction of science were accommodated and integrated in the decision-making process: not via legal restrictions and state-political jurisdiction but via the prescription of discursive standards and parameters for applications and decision-making processes.

Altogether, the crucial similarity between both countries was that, in both, it was taken for granted that science inevitably progresses and this assumption determined the normative background frame in the processes of problematisation and discourse stimulation. The *imperative of scientific progress* became the dominant frame in forming problematisations, such as those of 'the potential of concerns to block scientific development' or of 'serious diseases' in Great Britain, and those of 'restricting options or interests of concerned people' or of a conflict between a 'German *Sondermoral* and the freedom of research' in Germany. However, whereas techno-sceptical problematisations countering a scientific progress imperative also existed in Germany—at least in the beginning of the debates—this was not the case in Great Britain, at least not noticeably so. Nevertheless, in Germany the imperative of scientific progress likewise survived adversarial conflicts. This was most obvious in advisory reports on stem cell research issued by the National Ethics Council and the parliamentary Enquête Commission on Biomedicine. The reports referred to 'future developments of science' and excluded the option that science may not proceed from the outset. Chancellor Schröder was most explicit on this subject by declaring that 'no power in the world can stop knowledge' and that this was even less likely on a national scale. In Great Britain, on the other hand, the assumption of the inevitability of scientific 'progress' more evidently guided the search for problematisations from the beginning, including problematisation of 'serious diseases', whereby scientific development, i.e. 'therapeutic' cloning, was transformed into a solution and thus de-problematised. Scientific 'progress' was presented as the prerequisite for the well-being of the individual as well as of the nation: the commitment to societal 'progress' by means of scientific 'progress' remained stable and unquestionable. Therefore we find in both countries, although to varying extents, resignation towards being able to politically shape scientific development.

A second crucial similarity between the two countries was a growing focus on the 'status of the embryo' at the close of the debates and problematisation of 'different individual ethical perspectives on this status'. This focus not only underpinned the imperative of scientific progress—by excluding science or scientific developments from problematisation—but also helped to frame the issues at hand as matters for individual moral judgement and beliefs and not as a matter for politics to decide. This was most obviously the case in the British parliamentary debates, where general limits to scientific research were disqualified as an illegitimate 'imposition' on personal moral values and beliefs. In Germany, on the other hand, the focus on the 'status of the embryo' not only helped to integrate techno-sceptical voices into the dominant discourse but also to cleanse the debate of its adversarial character. Thereby the existence of a year-long, very conflict-ridden debate was denied, in which very different problematisations—including problematisation of scientific 'progress'—came to light. With the problematisation of 'different ethical viewpoints on the status of the embryo' in both countries, the imperative of scientific progress was overshadowed but certainly not overcome. Indeed, it remained the unquestionable background principle.

The frequent call for more or more rational public discourse was likewise the result of the imperative of scientific progress: Many quotes suggest that (especially techno-optimist) actors promoted more debate as a means for dealing with existing or anticipated opposition to or conflicts over scientific developments. Public engagement in discourse was to alleviate 'a deficit of trust' and to create an open environment in which scientific possibilities can be realized. Stimulation of discourse should, in fact, accompany ongoing scientific development, excluding, in principle, that it be challenged more generally. In this respect we see, especially in Great Britain, a transformation of the function of public debate in regard to public 'concerns': While public 'concerns' were initially depicted as deriving from misinformation, calling for more education by means of public debate, they were finally portrayed as 'different ethical viewpoints', which must be respected and integrated into public debate. Integration and respect, however, were not directed towards finding a definite, common answer, but rather towards preparing individuals to make a moral decision. Similarly, stimulating public discourse in Germany—an obvious example is the statement made by the National Ethics Council on embryonic stem cell research—was directed at generating a personal opinion on the new research rather than finding a mutual answer. Indeed, in both countries, finding a mutual solution to the 'ethical' conflict was interpreted as being impossible. Problematisations of '*irreconcilable* ethical perspectives on the status of the embryo' in both countries therefore underpinned calls for public engagement in discourse. However, this discourse accompanied and did not challenge scientific developments and was not directed at finding a political, mutual answer. The focus on the status of the embryo referred potential or actual conflicts and concerns over, opposition to, or rejection of new research to 'ethics' as an area separate from science and from politics—an area that was almost by definition outside of intervention by the sate. This opened up an endless discursive space for these conflicts and concerns whilst successfully shielding scientific development from more general societal and political scrutiny. Indeed, referring (actual or potential) conflicts to the area of 'ethics' signalised a depoliticisation of science in favour of liberal and individualistic market logic: Both countries were marked in different ways by the tendency to take 'ethical issues' into account by constructing them in terms of consumption: In Great Britain, 'ethics' meant that individuals could choose from a set of options produced by science according to their 'ethical' preferences on such issues as the donation of egg cells or using certain therapies. In Germany, on the other hand, the National Ethics Council offered a set of possible 'ethical' positions from which interested actors can choose accordingly. Participation in 'ethical' debates is therefore not actually participation in a political process that shapes scientific endeavour, but is directed at personal decision-making according to the personal (moral) convictions of an 'ethical' consumer. In the process neither public debate nor 'ethics' counter the imperative of scientific progress. Rather, the stimulation of (public) discourse, especially of a discourse in terms of 'ethics', responds to a conflict between the imperative of scientific progress, on the one hand, and the existence of concerns or possibility of opposition, on the other. This response, however, serves the stabilisation of the imperative rather than challenging it.

Abbreviations

ART	assisted reproductive technologies
Art.	Article
BÄK	*Bundesärztekammer* (German Federal Medical Association)
BBSRC	Biotechnology and Biological Sciences Research Council, Branch within the IPMS
BMA	British Medical Association
BMBF	*Bundesministerium für Bildung und Forschung* (German Federal Ministry of Education and Research)
BMFT	*Bundesministerium für Forschung und Technologie* (German Federal Ministry of Research and Technology)
BMG	*Bundesministerium der Gesundheit* (German Federal Health Ministry)
BMJ	*Bundesministerium der Justiz* (German Federal Ministry of Justice)
BT DR	*Bundestags-Drucksache* (German Parliamentary Proceedings)
CAHGE	Campaign Against Human Genetic Engineering
CDBI	*Comité Directeur pour la Bioéthique*
CMO	Chief Medical Officer
CNR	cell nuclear replacement
col.	column
CWG	Cloning Working Group (by the HFEA and HGAC)
DFG	*Deutsche Forschungsgemeinschaft* (German Research Association)
DoH	Department of Health
DTI	Department of Trade and Industry
EK DB	*Enquête-Kommission des Deutschen Bundestages* (Enquête Commission of the German Bundestag)
EK REM	*Enquête-Kommission des Deutschen Bundestages Recht und Ethik der Modernen Medizin* (Enquête Commission Law and Ethics of Modern Medicine)
ES cell	embryonic stem cell
EschG	*Embryonenschutzgesetz* (Embryo Protection Act)
FmedG	*Fortpflanzungsmedizingesetz* (Reproductive Medicine Act)

GeN	*Gen-ethisches Netzwerk* (Gene-ethics Network)
GenTG	*Gentechnologiegesetz* (Genetic Technology Act)
GfH	*Deutsche Gesellschaft für Humangenetik* (German Society for Human Genetics)
GG	*Grundgesetz* (German Basic Law)
GiD	*Gen-ethischer Informationsdienst* (Gen-ethics Information Service)
GIG	Genetic Interest Group
GMO	genetically modified organisms
HFE Act	Human Fertilisation and Embryology Act
HFEA	Human Fertilisation and Embryology Authority
HFE Bill	Human Fertilisation and Embryology Bill
HGAC	Human Genetics Advisory Commission
HGC	Human Genetics Commission
IMEW	*Institut Mensch, Ethik, Wissenschaft* (Institute of Humankind, Ethics, and Science)
IPMS	Institution of Professionals, Managers and Specialists
IVF	in-vitro fertilisation
MAFF	Ministry of Agriculture, Fisheries and Food
MP	Member of Parliament
MUL	*Medizinische Universität zu Lübeck* (Lübeck Medical School)
NER	*Nationaler Ethikrat* (German National Ethics Council)
OST	Office of Science and Technology
PGD	pre-implantation genetic diagnosis
PLP	*Plenarprotokoll des Deutschen Bundestages*, (Plenary Minutes of the German Parliament)
PND	prenatal diagnosis
PUS	Public Understanding of Science
RKI	Robert-Koch-Institute
STC	Science and Technology Committee
StGB	*Strafgesetzbuch* (German Penal Code)
ST&S	science, technology and society
StZG	*Gesetz zur Sicherung des Embryonenschutzes im Zusammenhang mit Einfuhr und Verwendung menschlicher embryonaler Stammzellen (Stammzellgesetz)*, (Law to Secure Embryo Protection with Regard to Imports and Use of Human Embryonic Stem cells, Stem Cell Act)
ZES	*Zentrale Ethikkommission für Stammzellforschung* (Central Ethics-Commission for Stem Cell Research)

References

Ach, Johann S.,Gerd Brudermüller and Christa Runtenberg (eds.) (1998). *Hello Dolly? Über das Klonen.* Frankfurt a.M.: Edition Suhrkamp.

Aretz, Hans-Jürgen (1999). *Kommunikation ohne Verständigung: das Scheitern des öffentlichen Diskurses über die Gentechnik und die Krise des Technokorporatismus in der Bundesrepublik Deutschland.* Frankfurt a.M.: Peter Lang Verlag.

Bacchi, Carol Lee (1990). *Same Difference: Feminism and Sexual Difference.* Sydney: Allen and Unwin.

— (1999). *Women, Policy and Politics. The Construction of Policy Problems.* London/ Thousand Oaks/New Delhi: Sage.

Bachrach, Peter and Morton Baratz (1963). Decisions and Nondecisions: An Analytical Framework. *American Political Science Review* 57 (3), 632–642.

Bahners, Patrik, Christian Geyer, and Christian Schwägerl (2001). Die Würde des Embryos ist unbezweifelbar. Gespräch mit Herta Däubler-Gmelin. In Christian Geyer (ed.). *Biopolitik. Die Positionen.* Frankfurt a.M.: Edition Suhrkamp, 121–139.

BÄK (2001). Bundesärztekammer. Diskussionsentwurf zu einer Richtlinie zur Präimplantationsdiagnostik. In Sigrid Graumann (ed.). *Die Genkontroverse. Grundpositionen.* Freiburg/Basle/Vienna: Herder Spektrum, 157–168.

— (2006). Bundesärztekammer. Muster-Richtlinie zur Durchführung der assistierten Reproduktion. Novelle 2006. *Deutsches Ärzteblatt.* 103, Vol. 20, 19. Mai 2006, A1392–A1403. Feb 8, 2007, http://www.bundesaerztekammer.de/down loads/Kuenstbefrucht_pdf.pdf.

Bandelow, Nils C. and Klaus Schubert (eds.) (2003). *Lehrbuch der Politikfeldanalyse.* Munich: Oldenbourg.

Bauer, Martin W., John Durant and George Gaskell (eds.) (1998). *Biotechnology in the Public Sphere: A European Sourcebook.* London: Science Museum.

BBSRC (1997). Institution of Professionals, Managers and Specialists (BBSRC Branch). Memorandum submitted by the Institution of Professionals, Managers and Specialists (BBSRC Branch) (6 March 1997) (CLE 4). In STC (ed.) (1997b). *Fifth Report of Sessions 1996–1997 – The Cloning of Animals from Adult Cells—Vol. II Minutes of Evidence and Appendices.* London: The Stationary Office, 34–41.

Beier, Henning M., K. Diedrich, H. Hepp, E. Nieschlag, B. Runnebaum, and W.B. Schill (1994). Gesetzliche Regelungen der menschlichen Fortpflanzung. Stellungnahme zum Musterentwurf eines Fortpflanzungsmedizingesetzes der Länder. *Fertilität* 10, 122–123.

Beier, Henning M. (1996). *Assistierte Reproduktion. Zum Zustand der Therapieverfahren in der Bundesrepublik Deutschland. Gutachten im Auftrag des Bundesministeriums für Gesundheit.* Bonn: Bundesministerium für Gesundheit.

Berg, Giselind (2002). Current Situation. In ReproKult (ed.) (2002a). *Reproductive Medicine and Genetic Engineering. Women Between Self-Determination and Societal Standardisation.* Bonn: Bundeszentrale für gesundheitliche Aufklärung, 13–17.

Berg, Giselind, Sigrid Graumann and Ingrid Schneider (2001). Fortschrittsgläubiger Rückfall in alte Fronten. *Freitag* 2001(3).

BMA (1997). British Medical Association. Memorandum submitted by the British Medical Association (10 March 1997) (CLE 6). In STC (ed.) (1997b). *Fifth Report of Sessions 1996–1997 – The Cloning of Animals from Adult Cells—Vol. II Minutes of Evidence and Appendices.* London: The Stationary Office, 41.

BMFT (1984). Der Bundesminister für Forschung und Technologie. *Ethische und rechtliche Problem der Anwendung zellbiologischer und genetischer Methoden am Menschen. Dokumentation des Fachgesprächs im Bundesministerium für Forschung und Technologie.* Munich: J.Schweitzer Verlag.

— (1985). Der Bundesminister für Forschung und Technologie. *In-vitro-Fertilisation, Genomanalyse und Gentherapie. Bericht der gemeinsamen Arbeitsgruppe des Bundesministers für Forschung und Technologie und des Bundesministers der Justiz.* Munich: J.Schweitzer Verlag.

BMG (2001). Bundesministerium der Gesundheit. *Fortpflanzungsmedizin in Deutschland. Wissenschaftliches Symposium des Bundesministeriums für Gesundheit in Zusammenarbeit mit dem Robert Koch-Institut, 24.–26. Mai 2000.* Berlin: Nomos Verlag.

Bradish, Paula, Erika Feyerabend and Ute Winkler (eds.) (1989). *Frauen gegen Gen- und Reproduktionstechnologien. Beiträge vom 2. Bundesweiten Kongress, Frankfurt, 28. – 30.10.1988.* Munich: Frauenoffensive.

Braun, Kathrin (1998). Zivilisation durch Dehumanisierung? Menschenrechtsschutz und internationales Bioethikregime. *Österreichische Zeitschrift für Politikwissenschaft* 1998 (1), 63–77.

— (2000a). Kann man über alles reden? Bioethik und demokratischer Diskurs. In Christian Mürner, Adelheid Schmitz, and Udo Sierk (eds.) (2000). *Schöne Heile Welt? Biomedizin und Normierung des Menschen.* Hamburg/Berlin: Schwarze Risse Rote Strasse, 175–187.

— (2000b). *Menschenwürde und Biomedizin. Zum Philosophischen Diskurs der Bioethik.* Frankfurt a.M./New York: Campus.

— (2002). Deliberation, Democracy, and Bioethics: How to Democratise the Politics of Biomedicine. Paper presented at the ECPR Joint Session of Workshops, Workshop 18: *The Politics of Biomedicine. Explaining and Evaluating Policy Design*, March 22–27, in Turin, Italy.

— (2003). Eine feministische Verteidigung des Menschenwürdeschutzes für menschliche Embryonen. In Sigrid Graumann and Ingrid Schneider (eds.) (2003), *Verkörperte Technik—Entkörperte Frau. Biopolitik und Geschlecht.* Frankfurt a.M./New York: Campus.

— (2005). Not Just for Experts. The Public Debate about Reprogenetics in Germany. *Hastings Center Report* 35 (3, 2005), 42–49.

Braun, Kathrin and Svea Luise Herrmann (2000). If Discourse is the Solution—What is the Problem? Paper presented at the ECPR Joint Session of Workshops, Workshop 9: *Policy, Discourse and Institutional Reform*, April 6–11, in Grenoble, France. May 20, 2008, http://www.essex.ac.uk/ECPR/events/joint sessions/paperarchive/grenoble/ws9/braun.pdf.

Braun, Kathrin, Svea Luise Herrmann, and Sabine Könninger (2002). *Deliberative Modelle als Mittel der Demokratisierung von Bioethik- und Biomedizinpolitik.* Gutachten im Auftrag der Brandenburgischen Akademie der Wissenschaften (BBAW). May 20, 2004, http://www.sciencepolicystudies.de/dok/expertise-braun.pdf.

Braun, Kathrin, Svea Luise Herrmann, Sabine Könninger, and Alfred Moore (2008). Die Sprache der Ethik und die Politik des richtigen Sprechens. Ethikregime in Deutschland, Frankreich und Großbritannien. In Renate Mayntz, Friedhelm Neidhardt, Peter Weingart, and Ulrich Wengenroth (eds.) (2008). *Wissensproduktion und Wissenstransfer. Wissen im Spannungsfeld von Wissenschaft, Politik und Öffentlichkeit.* Bielefeld: transcipt, 221–242.

Braun, Kathrin, Svea Luise Herrmann, Sabine Könninger, and Alfred Moore (2009). Bioethik in der Politik: Vom technokratischen Model zur Politik des richtigen Sprechens. *Aus Politik und Zeitgeschichte* 8/2009, 40–46.

Campbell, Colin and Ruth Deech (1998). Consulting on Cloning. *SPA* Spring 1998, 31–33.

Carr, Susan and Les Levidow (1997). How Biotechnology Regulation Sets a Risk/Ethics Boundary. *Agriculture and Human Values*, 14, 29–43.

Deutscher Bundestag (2001). *Zweiter Zwischenbericht der Enquête-Kommission Recht und Ethik der modernen Medizin. Teilbericht Stammzellforschung, BT DR 14/7546.* Berlin: Deutscher Bundestag.

— (2002). *Schlussbericht der Enquête-Kommission 'Recht und Ethik der modernen Medizin'.* Berlin: Deutscher Bundestag.

DFG (1996a). Deutsche Forschungsgemeinschaft. Forschungsfreiheit—ein Plädoyer der Deutschen Forschungsgemeinschaft (DFG) für bessere Rahmenbedingungen der Forschung in Deutschland. *Der Frauenarzt* 37 (7), 1007–1023.

— (1996b). Deutsche Forschungsgemeinschaft. *Forschungsfreiheit: ein Plädoyer der Deutschen Forschungsgemeinschaft für bessere Rahmenbedingungen der Forschung in Deutschland.* Weinheim: VCH.

— (2001). Deutsche Forschungsgemeinschaft. Die Empfehlungen der Deutschen Forschungsgemeinschaft zur Forschung mit menschlichen Stammzellen. In

Sigrid Graumann (ed.) (2001). *Die Genkontroverse. Grundpositionen.* Freiburg/Basle/Vienna: Herder Spektrum, 107–114.

Die Grünen im Bundestag and Arbeitskreis Frauenpolitik e.V. (eds.) (1986). *Frauen gegen Gen- und Reproduktionstechnik. Dokumentation zum Kongress vom 19.–21.4.1984 in Bonn.* Cologne: Kölner Volksblatt Verlag.

DoH (1997). Department of Health. Memorandum submitted by the Department of Health (5 March 1997) (CLE 3). In STC (ed.) (1997b). *Fifth Report of Sessions 1996–1997—The Cloning of Animals from Adult Cells—Vol. II Minutes of Evidence and Appendices.* London: The Stationary Office, 31–34.

— (1999). Department of Health. *Government Response to the Report by the Human Fertilisation and Embryology Authority and the Human Genetics Advisory Commission on Cloning Issues in Reproduction, Science and Medicine.* London: The Stationary Office.

— (2000a). Department of Health. *Stem Cell Research: Medical Progress with Responsibility. A Report from the Chief Medical Officer's Expert Group Reviewing the Potential of Developments in Stem Cell Research and Cell Nuclear Replacement to Benefit Human Health.* London: The Stationary Office.

— (2000b). Department of Health. *Government Response to the Recommendations made in the Chief Medical Officer's Expert Group Report 'Stem Cell Research: Medical Progress with Responsibility'.* London: The Stationary Office.

Dolata, Ulrich (1996). *Politische Ökonomie und Gentechnik. Konzernstrategien, Forschungsprogramme, Technologiewettläufe.* Berlin: Edition Sigma.

Dreyfus, Hubert L. and Paul Rabinow (1982). *Michel Foucault. Beyond Structuralism and Hermeneutics.* Chicago: University of Chicago Press.

Dryzek, John (1993). Policy Analysis and Planning: From Science to Argument. In Frank Fischer and John Forester (eds.) (1993a). *The Argumentative Turn in Policy Analysis and Planning.* Durham/London: Duke University Press, 213–232.

— (2002). *Deliberative Democracy and Beyond: Liberals, Critics, Contestations.* Oxford: Oxford University Press.

DTI and OST (1997). Department of Trade and Industry and Office of Science and Technology. *The Cloning of Animals from Adult Cells—Government Response to the Fifth Report of the House of Commons' Select Committee on Science and Technology, 1996–97 Session.* London: The Stationery Office.

Duden, Barbara (1993). *Disembodying Women: Perspective on Pregnancy and the Unborn.* Cambridge: Harvard University Press.

— (2002). Die höchstrichterliche Anerkennung des körperlosen Menschen: ein deutscher Sonderweg. In Barbara Duden (2002), *Die Gene im Kopf—der Fötus im Bauch. Historisches zum Frauenkörper.* Hanover: Offizin, 234–245.

Durant, John and Simon Joss (1995). Introduction. In John Durant and Simon Joss (eds.) (1995). *Public Participation in Science. The Role of Consensus Conferences in Europe.* London: The Science Museum, 9–13.

Duster, Troy (1990). *Backdoor to Eugenics.* London/New York: Routledge

Edelman, Murray (1988). *Constructing the Political Spectacle.* Chicago: University of Chicago Press.

EK DB, Wolf-Michael Catenhusen and Hanna Neumeister (1987). *Chancen und Risiken der Gentechnologie. Bericht der Enquête-Kommission "Chancen und Risiken der Gentechnologie" des 10. Deutschen Bundestages. (BT Drucksache 10/6775.).* Bonn: Referat für Öffentlichkeitsarbeit.

Emmrich, Michael (1997). *Der vermessene Mensch. Aufbruch ins Gen-Zeitalter.* Berlin: Aufbau Taschenbuch Verlag.

— (1999). Abschied vom Nürnberger Kodex. Die Bioethik-Konvention: Der Triumph des utilitaristischen Denkens. In Michael Emmrich (ed.) (1999). *Im Zeitalter der Bio-Macht. 25 Jahre Gentechnik—eine kritische Bilanz.* Frankfurt a.M.: Mabuse-Verlag, 341–348.

Erklärung [1989] (1991). Erklärung deutscher Philosophen zur sog. "Singer-Affaire". In Rainer Hegselmann and Reinhard Merkel (eds.) (1991). *Zur Debatte über Euthanasie: Beiträge und Stellungnahmen.* Frankfurt a.M.: Suhrkamp, 327–330.

Fairclough, Norman (1992). *Discourse and Social Change.* Cambridge: Polity Press.

Felt, Ulrike, Helga Nowotny and Klaus Taschner (1995). *Wissenschaftsforschung. Eine Einführung.* Frankfurt a.M./New York: Campus.

Fiorino, D. J. (1989). Technical and Democratic Values in Risk Analysis. *Risk Analysis* 9 (3), 293–299.

Fischer, Andrea (2001a). Schlusswort. In BMG (ed.) (2001). Bundesministerium der Gesundheit. *Fortpflanzungsmedizin in Deutschland. Wissenschaftliches Symposium des Bundesministeriums für Gesundheit in Zusammenarbeit mit dem Robert Koch-Institut.* Berlin: Nomos Verlag, 326–329.

— (2001b). Wir sind für Gentechnik. Aber nicht auf Kosten des Embryos. In Christian Geyer (ed.) (2001). *Biopolitik. Die Positionen.* Frankfurt a.M.: Edition Suhrkamp, 151–154.

Fischer, Frank (1999). Technological Deliberation in a Democratic Society: The Case for Participatory Inquiry. *Science and Public Policy* 26 (5), 294–302.

— (2000). *Citizens, Experts, and the Environment.* Durham/London: Duke University Press.

— (2003a). Beyond Empiricism: Policy Analysis as Deliberative Practice. In Marten Hajer and Hendrik Wagenaar (eds.) (2003). *Deliberative Policy Analysis. Understanding Governance in the Network Society.* Cambridge: Cambridge University Press, 209–227.

— (2003b). *Reframing Public Policy. Discourse Politics and Deliberative Practice.* Oxford: University Press.

Fischer, Frank and John Forester (eds.) (1993a). *The Argumentative Turn in Policy Analysis and Planning.* Durham/London: Duke University Press.

Fischer, Frank and John Forester (1993b). Editors' Introduction. In Frank Fischer and John Forester (eds.) (1993a). *The Argumentative Turn in Policy Analysis and Planning.* Durham/London: Duke University Press, 1–17.

Forbes, Ian (2004). Making a Crisis out of a Drama: The Political Analysis of BSE–Policy Making in the UK. *Political Studies* 52, 342–357.
Foucault, Michel (1974). The Order of Discourse. In Robert Young (ed.) (1981). *Untying the Text: A Poststructuralist Reader.* London: Routledge, 51–78.
— (1978). Die Machtverhältnisse durchziehen das Körperinnere. In *Dispositive der Macht. Michel Foucault über Sexualität, Wissen und Wahrheit.* Berlin: Merve, 104–117.
— (1980). Truth and Power. In Colin Gordon (ed.) (1980). *Michel Foucault: Power/Knowledge: Selected Interview and Other Writings, 1972–1977.* New York: Pantheon Books, 109–133.
— (1989). *The Archaeology of Knowledge.* London: Routledge.
— (1994). The Subject and Power. In James D. Faubion (ed.) (1994). *Power. Essential Works of Foucault 1954 – 1984. Vol. 3.* London: Penguin Books, 326–348.
— (1998). *The Will to Knowledge. The History of Sexuality I.* London: Penguin Books.
Frankenberg, Günter (2005). *Grundgesetz.* Frankfurt a.M.: Fischer.
Franklin, Sarah (1993). Making Representations. The Parliamentary Debates on the Human Fertilisation and Embryology Act. In Jeanette Edwards, Sarah Franklin, Eric Hirsch, Frances Price and Marilyn Strathern (eds.) (1993). *Technologies of Procreation: Kinship in the Age of Assisted Conception.* London: Routledge, 127–170.
— (1997). *Embodied Progress. A Cultural Account of Assisted Conception.* London/New York: Routledge.
— (1999a). *Animal Models: An Anthropologist Considers Dolly.* Published by the Department of Sociology, Lancaster University. Aug 12, 2002, http://www.comp.lancs.ac.uk/sociology/soc022sf.html.
— (1999b). Dead Embryos: Feminism in Suspension. In Lynn M. Morgan and Meredith W. Michaels (eds.) (1999). *Fetal Subjects, Feminist Positions.* Pennsylvania: University of Pennsylvania Press, 61–82.
Fraser, Nancy (1989a). Struggle over Needs: Outline of a Socialist-Feminist Critical Theory of the Late Capitalist Political Culture. In Nancy Fraser (1989), *Unruly Practices. Power, Discourse, and Gender in Contemporary Social Theory.* Minneapolis: University of Minnesota Press, 161–190.
— (1989b). Women, Welfare, and the Politics of Need Interpretation. In Nancy Fraser (1989), *Unruly Practices. Power, Discourse, and Gender in Contemporary Social Theory.* Minneapolis: University of Minnesota Press, 144–160.
— (1997a). Rethinking the Public Sphere: A Contribution to the Critique of Actually Existing Democracy. In Nancy Fraser (1997). *Justice Interruptus. Critical Reflections on the "Postsocialist" Condition.* New York/London: Routledge, 69–98.
— (1997b). Sex, Lies, and the Public Sphere: Reflections on the Confirmation of Clarence Thomas. In Nancy Fraser (1997). *Justice Interruptus. Critical Reflections on the "Postsocialist" Condition.* New York/London: Routledge, 99–120.

Friedrich-Ebert-Stiftung (1997). *Braucht Deutschland eine Bundes-Ethik-Kommission. Dokumentation des Expertengesprächs Gentechnik am 11. März 1997 in Bonn.* Bonn: Friedrich-Ebert-Stiftung.

Frommel, Monika (2001). Status des Embryos: Juristische Aspekte. In BMG (ed.) (2001). Bundesministerium der Gesundheit. *Fortpflanzungsmedizin in Deutschland. Wissenschaftliches Symposium des Bundesministeriums für Gesundheit in Zusammenarbeit mit dem Robert Koch-Institut.* Berlin: Nomos Verlag, 67–75.

Fuchs, Michael (2001). *Internationaler Überblick zu Verfahren der Entscheidungsfindung bei ethischem Dissens. Gutachten für die Enquête-Kommission "Recht und Ethik der modernen Medizin" des Deutschen Bundestages. Arbeitsunterlage 14/55.* Jan 1, 2002, http://www.bundestag.de/gremien/medi/medi_ext.html.

Fuchs, Ursel (1998a). Bürger gegen Bio-Ethik. Internationale Initiative gegen die geplante Bio-Ethik-Konvention. In Michael Wunder and Therese Neuer-Miebach (eds.) (1998a). *Bio-Ethik und die Zukunft der Medizin.* Bonn: Psychiatrie-Verlag, 165–166.

— (1998b). Experten entscheiden selbst. Unter sich. Und über uns. Die Bio-Ethik-Konvention geht alle an. In Michael Wunder and Therese Neuer-Miebach (eds.) (1998a). *Bio-Ethik und die Zukunft der Medizin.* Bonn: Psychiatrie-Verlag, 130–138.

— (1999). Die Ethik der Bio-Macht. Bioethik oder: Tabubrüche hinter verschlossenen Türen. In Michael Emmrich (ed.) (1999). *Im Zeitalter der Bio-Macht. 25 Jahre Gentechnik–eine kritische Bilanz.* Frankfurt a.M.: Mabuse-Verlag, 261–273.

Gaskell, George, Martin W. Bauer and John Durant (1998). The Representation of Biotechnology: Policy, Media and Public Perception. In John Durant, Martin W. Bauer, and George Gaskell (eds.) (1998). *Biotechnology in the Public Sphere: A European Sourcebook*, London: Science Museum.

Gaskell, George, Martin W. Bauer and John Durant (2000). Biotechnology and the European Public. *Nature Biotechnology* 18, 935–938.

Gaskell, George and Martin W. Bauer (eds.) (2001). *Biotechnology 1996–2000—the Years of Controversy.* London: Science Museum.

Gaskell, George, Martin W. Bauer, Jonathan Jackson, Susan Howard, and Nicola Lindsey (2003). *Ambivalent GM Nation? Public Attitides to Biotechnology in the UK 1991–2002. "Life Sciences in European Society" Report.* London: London School of Economics and Political Sciences.

Geisler, Linus (2002). Enquête-Kommissionen und Biopolitik. Erfahrungen und Überlegungen am Beispiel der Enquête-Kommission Recht und Ethik der modernen Medizin. *Zeitschrift für Biopolitik* 3 (1), 23–30.

German National Ethics Council (2001). *The Import of Human Embryonic Stem Cells. Opinion.* Berlin: Nationaler Ethikrat Deutschland.

Geyer, Christian (ed.) (2001). *Biopolitik. Die Positionen.* Frankfurt a.M.: Edition Suhrkamp.

GfH (1995). Kommission für Öffentlichkeitsarbeit und ethische Fragen der Gesellschaft für Humangenetik e.V. und Berufsverband Medizinische Genetik e.V.. Stellungnahme zur Präimplantationsdiagnostik. *Medizinische Genetik* 7, 420.

GID (2001). Die Kehrseite der Macht. Blick ins Medienecho zu des Kanzler 'Nationalem Ethikrat'. *Gen-ethischer Informationsdienst GID* 146, 40–41.

Gieryn, Thomas F. (1983). Boundary Work and the Demarcation of Science from Non-Science: Strains and Interests in Professional Ideologies of Scientists. *American Sociological Review* 48 (6), 781–795.

Gill, Bernhard (1991a). Die Bundestags-Enquête zur Gentechnik—Lehrstück einer politischen Nicht-Entscheidung. *Leviathan* 3, 412–434.

— (1991b). *Gentechnik ohne Politik. Wie die Brisanz der synthetischen Biologie von wissenschaftlichen Institutionen, Ethik- und anderen Kommissionen systematisch verdrängt wird.* Frankfurt a.M./New York: Campus.

Goffman, Erving (1974). *Frame Analysis: An Essay on the Organisation of Experience.* Cambridge: Harvard University Press.

Gottweis, Herbert (1998). *Governing Molecules. The Discursive Politics of Genetic Engineering in Europe and the US.* Cambridge: The MIT Press.

Grafenecker Erklärung [1995] (1998). Granfenecker Erklärung zur Bio-Ethik. In Michael Wunder and Therese Neuer-Miebach (eds.) (1998a). *Bio-Ethik und die Zukunft der Medizin.* Bonn: Psychiartrie-Verlag, 182–195.

Graumann, Sigrid (2000). Gen-Check vor der Schwangerschaft. *Gen-ethischer Informationsdienst GID* 139, 13–16.

— (2001). Die Genkontroverse. In Sigrid Graumann (ed.) (2001). *Die Genkontroverse: Grundpositionen.* Freiburg/Basle: Herder, 7–13.

— (2002a). Liberation or Disempowerment of Women—The Social Consequences of Reproductive Medicine and Genetic Diagnosis from the Standpoint of Women's Policy. In ReproKult (ed.) (2002a). *Reproductive Medicine and Genetic Engineering. Women Between Self-Determination and Societal Standardisation.* Bonn: Bundeszentrale für gesundheitliche Aufklärung, 111–115.

— (2002b). *Situation der Medienberichterstattung zu aktuellen Entwicklungen in der Biomedizin und ihren ethischen Fragen. Gutachten für die AG 'Bioethik und Wissenschaftskommunikation' am Max-Delbrück-Centrum für Molekulare Medizin Berlin.* Oct 20, 2002, http://www.bioethik-diskurs.de/documents/wissensdatenbank/gutachten/Download-Dokumente/Graumann-Gutachten

Grove-White, Robin (2006). Britain's GM Crop Controversies: the AEBC and the Negotiation of "Uncertainty". *Community Genet* 9 (2006), 170–177.

Habermas, Jürgen (1995). Wahrheitstheorien. In Jürgen Habermas (1995). *Vorstudien und Ergänzungen zur Theorie des kommunikativen Handelns.* Frankfurt a.M.: Suhrkamp, 127–183.

Hajer, Marten (1993). Discourse Coalitions and the Institutionalisation of Practice: The Case of Acid Rain in Britain. In Frank Fischer and John Forester (eds.)

(1993a). *The Argumentative Turn in Policy Analysis and Planning*. Durham/London: Duke University Press, 43–76.

Hajer, Marten and Hendrik Wagenaar (2003). Introduction. In Marten Hajer and Hendrik Wagenaar (eds.) (2003). *Deliberative Policy Analysis. Understanding Governance in the Network Society*. Cambridge: Cambridge University Press, 1–30.

Herrmann, Svea Luise (1999). Bioethics Education. Bioethik und Normalisierung. In Kathrin Braun (ed.) (1999). *"Life" is a Battlefield. Aspekte der Biomacht*. Hanover: Offizin, 69–98.

— (2002). What is the Problem? Policy Discourses on Reprogenetics in Germany since the mid-1990s. Paper presented at the ECPR Joint Session of Workshops, Workshop 18: *The Politics of Biomedicine. Explaining and Evaluating Policy Design*, March 22–27, in Turin, Italy, http://www.essex.ac.uk/ ECPR/events/jointsessions/paperarchive/turin/ws18/Herrmann.pdf.

— (2003). Deregulation via Regulation. On the Moralisation and Naturalisation of Embryonic Stem Cell Research in the British Parliamentary Debates of 2000/2001. *Österreichische Zeitschrift für Politikwissenschaft* 32, (2003/2), 149–161.

— (2006). Bioethicspolicies in Germany: From Conflict Prevention to Conflict Management. Paper presented at the Conference *The Politics of Ethics and the Crisis of Government. Contested Technologies, the Language of Ethics and the Transformation of Governance in Europe and North America*, May 25–26 at the University of Washington, Seattle, USA

— (2008). Ethik als diskursives Schmiermittel. *Gen-ethischer Informationsdienst GiD* 188, June 2008, 56–59.

Herrmann, Svea Luise and Sabine Könninger (2008). "… but you can not influence the direction of your thinking." Guiding Self-Government in Bioethics Policy Discourse. In Barbara Katz Rothman, Elizabeth M. Armstrong, and Rebecca Tiger (eds.) (2008). *Bioethical Issues: Sociological Perspectives*. Amsterdam/Oxford: Elsevier, 207–225.

Herrmann, Svea Luise and Margaretha Kurmann (2002). Foreword. In ReproKult (ed.) (2002a). *Reproductive Medicine and Genetic Engineering. Women Between Self-Determination and Societal Standardisation*. Bonn: Bundeszentrale für gesundheitliche Aufklärung, 6.

Herwig, Eckart and BMFT (eds.) (1980). *Chancen und Gefahren der Genforschung: Protokolle und Materialien zur Anhörung des Bundesministers für Forschung und Technologie in Bonn, 19. – 21. September 1979*. Munich: Oldenbourg.

HFEA and HGAC (1998a). Human Fertilisation and Embryology Authority and Human Genetics Advisory Commission. *Cloning Issues in Reproduction, Science and Medicine—A Consultation Paper*. London: Department of Health. Feb 11, 2003, http://www.doh.gov.uk/hgac/papers/papers_c.htm.

HFEA and HGAC (1998b). Human Fertilisation and Embryology Authority and Human Genetics Advisory Commission. *Cloning Issues in Reproduction, Science and Medicine—Report*. London: Department of Health. Feb 11, 2003, http://www.doh.gov.uk/hgac/papers/papers_d.htm.

Höfling, Wolfram (1998). Das "Menschenrechtsübereinkommen zur Bio-Medizin" und die Grund- und Menschenrechte. In Michael Wunder and Therese Neuer-Miebach (eds.) (1998a). *Bio-Ethik und die Zukunft der Medizin*. Bonn: Psychiatrie-Verlag, 72–86.

Hülsmann, Christoph and Hans-Georg Koch (1990). Bundesrepublik Deutschland. In Albin Eser, Hans-Georg Koch, and Thomas Wiesenbart (eds.) (1990). *Regelungen der Fortpflanzungsmedizin und Humangenetik 1. Eine internationale Dokumentation gesetzlicher und berufsständischer Rechtsquellen*. Frankfurt a.M./New York: Campus, 29–155.

Ingram, Helen and Anne Larason Schneider (1997). *Policy Design for Democracy*. Lawrence/Kansas: University Press of Kansas.

Irwin, Alan (1995). *Citizen Science: A Study of People, Expertise and Sustainable Development*. London: Routledge.

— (2001). Constructing the Scientific Citizen: Science and Democracy in the Biosciences. *Public Understanding of Science* 10 (1), 1–18.

— (2006). The Politics of Talk. Coming to Terms with the "New" Scientific Governance. *Social Studies of Science* 36 (2), 299–320.

Irwin, Alan and Mike Michael (2003). *Science, Social Theory and Public Knowledge*. Maidenhead, Phil.: Open University Press.

Irwin, Alan and Brian Wynne (2003). *Misunderstanding Science? The Public Reconstruction of Science and Technology*. Cambridge: Cambridge University Press.

Jasanoff, Sheila (1994). *The Fifth Branch. Science Advisers as Policymakers*. Cambridge, MA: Harvard University Press.

— (2005). *Designs on Nature: Science and Democracy in Europe and the United States*. Princeton, NJ: Princeton Univ. Press.

Jordan, Isabella (1999). Kontinuität oder Bruch. Von der Vernichtung "lebensunwerten Lebens" zur "Neuen Euthanasie-"Diskussion. In Kathrin Braun (ed.) (1999). *"Life" is a battlefield. Aspekte der Biomacht*. Hanover: Offizin, 41–68.

Kerr, Anne (2003). Rights and Responsibility in the New Genetics Era. *Critical Social Policy* 23 (2003), 208–226.

Kitzinger, Jenny and Clare Williams (2005). Forecasting Futures: Legitimising Hope and Calming Fears in the Embryo Stem Cell Debate. *Social Science and Medicine* 61 (2005), 731–740.

Klüver, Lars et al. (2000). *EUROPTA. European Participatory Technology Assessment. Participatory Methods in Technology Assessment and Technology Decision-Making*. Oct 18, 2001, http://www.tekno.dk/europta.

Kollek, Regine (1999). Wegen der ethischen Brisanz nicht akzeptabel. *Gen-ethischer Informationsdienst GID* 131, 14–16.

— (2000). *Präimplantationsdiagnostik. Embryonenselektion, weibliche Autonomie und Recht*. Tübingen/Basle: Francke Verlag.

— (2001). Diskussion der Leitfrage 5, Statement der Podiumsredner: Regine Kollek. In BMG (ed.) (2001). Bundesministerium der Gesundheit. *Fortpflanzungsmedizin in Deutschland. Wissenschaftliches Symposium des Bundesministeriums für Gesundheit in Zusammenarbeit mit dem Robert Koch-Institut.* Berlin: Nomos Verlag, 440–441.

Lasswell, Harold D. (1970). The Emerging Conception of the Policy Sciences. *Policy Sciences* 1 (1970), 3–14.

Lindsey, Nicola, Mercy Wambui Kamara, Erlin Jelsoe and Alre Thing Mortensen (2001). Changing Frames: The Emergence of Ethics in European Policy and Biotechnology. *Politeia*, 17 (63), 80–93.

MacDonald, Christine (2003). The Value of Discourse Analysis as a Methodological Tool for Understanding a Land Reform Program. *Policy Sciences* 36 (2003), 151–173.

MacKinnon, Catherine A. (1989). *Abortion on Public and Private. Toward a Feminist Theory of the State.* Cambridge: Harvard University Press.

MAFF (1997). Ministry of Agriculture, Fisheries and Food. Memorandum submitted by the Ministry of Agriculture, Fisheries and Food (5 March 1997) (CLE 8). In STC (ed.) (1997b). *Fifth Report of Sessions 1996–1997—The Cloning of Animals from Adult Cells—Vol. II Minutes of Evidence and Appendices.* London: The Stationary Office, 47–48.

Majone, Giandomenico (1989). *Evidence, Argument and Persuasion in the Policy Process.* New Haven, CT: Yale University Press.

Markl, Hubert (2001). Freiheit, Verantwortung, Menschenwürde. Warum Lebenswissenschaften mehr sind als Biologie. In Christian Geyer (2001). *Biopolitik. Die Positionen.* Frankfurt a.M.: Suhrkamp Verlag, 177–193.

Marx Ferree, Myra, William Anthony Gamson, Jürgen Gerhards, and Dieter Rucht (2002). *Shaping Abortion Discourse. Democracy and the Public Sphere in Germany and the United States.* Cambridge: University Press.

Memmi, Dominique (2005). La Bio-politique déléguée. Paper presented at the Colloque *Le politique vue avec Foucault*, January 7–8, in Paris, France.

Merkel, Reinhard (2001). Rechte für Embryonen? Die Menschenwürde läßt sich nicht allein auf die biologische Zugehörigkeit zur Menschheit begründen. In Christian Geyer (ed.) (2001). *Biopolitik. Die Positionen.* Frankfurt a.M.: Edition Suhrkamp, 51–64.

Mildenberger, Elke H. (2002). Why Unwanted Pregnancies, Embryo Selection and Embryo Research must Generally be Treated Differently. In ReproKult (ed.) (2002a). *Reproductive Medicine and Genetic Engineering. Women Between Self-Determination and Societal Standardisation.* Bonn: Bundeszentrale für gesundheitliche Aufklärung, 115–119.

Morgan, Derek and Robert G. Lee (1991). *Human Fertilisation and Embryology Act 1990: Abortion and Embryo Research, the New Law.* London: Blackstone.

Mulkay, Michael (1993). Rhetoric of Hope and Fear in the Great Embryo Debate. *Social Studies of Science* 23, 721–742.

— (1994). The Triumph of the Pre-Embryo: Interpretations of the Human Embryo in Parliamentary Debates over Embryo Research. *Social Studies of Science* 24, 611–639.

— (1997). *The Embryo Research Debate.* Cambridge: Cambridge University Press.

Neidert, Rudolf (1998). Brauchen wir ein Fortpflanzungsmedizingesetz? *Medizinrecht* 16 (8), 347–353.

Nuffield Council on Bioethics (2000). *Stem Cell Therapy: The Ethical Issues.* April 2000. London: Nuffield Council on Bioethics.

Oehmichen, Manfred (1999a). Präimplantationsdiagnostik: Antrag und Entscheidungsfindung der Ethikkommission Lübeck. *Rechtsmedizin* 9 (1999), 107–111.

— (1999b). Votum der Ethikkommission der Medizinischen Universität zu Lübeck zur Präimplantationsdiagnostik. *Ethik in der Medizin* 11 (Supplement 1), 16–22.

OST (1997). Office of Science and Technology . Memorandum submitted by the Office of Science and Technology (10 March 1997) (CLE 7). In STC (ed.) (1997b). *Fifth Report of Sessions 1996–1997 – The Cloning of Animals from Adult Cells – Vol. II Minutes of Evidence and Appendices.* London: The Stationary Office, 42–47.

Palfner, Sonja (2003). *'Gene'. Diskurse und Politik in den 80er Jahren.* Unpublished Thesis at Otto-Suhr-Institut für Politikwissenschaft, April 2003, Berlin.

Parry, Sarah (2003). The Politics of Cloning: Mapping the Rhetorical Convergence of Embryos and Stem Cells in Parliamentary Debates. *New Genetics and Society* 22 (2), 145–168.

Parsons, Wayne (1999). *Public Policy. An Introduction to the Theory and Practice of Policy Analysis.* Cheltenham/Northampton: Edward Elgar.

Paul, Jobst (1998). *Die Bioethik- und Grundwerte-Debatte in Deutschland (1993–1998). Eine Dokumentation.* Duisburg: Diss.

Plough, A. and S. Krimsky (1987). The Emergence of Risk Communication Studies: Social and Political Context. *Science, Technology, and Human Values* 12, 4–10.

Prätorius, Ina (1985). Biotechnologien und Ethik – Fünf Thesen. In Die Grünen im Bundestag and AK Frauenpolitik e.V. (eds.) (1985). *Frauen gegen Gen- und Reproduktionstechnik. Dokumentation zum Kongress vom 19.–21.4.1984 in Bonn.* Cologne: Kölner Volksblatt Verlag, 92–94.

Radkau, Joachim (1988). Hiroshima und Asilomar. Die Inszenierung des Diskurses über die Gentechnik vor dem Hintergrund der Kernenergie-Kontroverse. *Geschichte und Gesellschaft* 14, 329–363.

Rau, Johannes (2001). Wird alles gut? Für einen Fortschritt nach menschlichem Maß. In Sigrid Graumann (ed.) (2001). *Die Genkontroverse. Grundpositionen.* Freiburg: Herder, 14–29.

Reimann, Carola (2000). Das Dilemma der Gentechnologie bleibt. Einstweilen könnte ein Nationaler Ethikrat der Forschung helfen. *Berliner Republik* 4 (2000), 10–13 .

Rein, Martin and Donald Schön (1993). Reframing Policy Discourse. In Frank Fischer and John Forester (eds.) (1993a). *The Argumentative Turn in Policy Analysis and Planning.* Durham/London: Duke University Press, 145–166.

Rein, Martin and Donald Schön (1994). *Frame Reflection.* New York: Basic Books.

Renesse, Margot v. (2000). Werden und Sterben. Die Arbeit der Enquête REM hat begonnen. *Berliner Republik* 4 (2000), 4–6.

ReproKult (2002a). *Reproductive Medicine and Genetic Engineering. Women Between Self-Determination and Social Standardisation. Proceedings of the Conference 15.–17. Nov. 2001, Berlin, Haus am Köllnischen Park.* Bonn: Bundeszentrale für gesundheitliche Aufklärung.

— (2002b). Position on Embryo and Embryonic Stem Cell Research. In ReproKult (ed.) (2002a). *Reproductive Medicine and Genetic Engineering. Women Between Self-Determination and Societal Standardisation.* Bonn: Bundeszentrale für gesundheitliche Aufklärung, 137–138.

— (2002c). Position on Preimplantation Genetic Diagnosis (PGD). In ReproKult (ed.) (2002a). *Reproductive Medicine and Genetic Engineering. Women Between Self-Determination and Societal Standardisation.* Bonn: Bundeszentrale für gesundheitliche Aufklärung, 135–136.

— (2002d). Human is Human. In ReproKult (ed.) (2002a). *Reproductive Medicine and Genetic Engineering. Women Between Self-Determination and Societal Standardisation.* Bonn: Bundeszentrale für gesundheitliche Aufklärung, 130–132.

Resolution (1986). Resolution. In Die Grünen im Bundestag and AK Frauenpolitik e.V. (eds.) (1986). *Frauen gegen Gen- und Reproduktionstechnik. Dokumentation zum Kongress vom 19.–21.4.1984 in Bonn.* Cologne: Kölner Volksblatt Verlag, 16–19.

Richardt, Nicole (2003). A Comparative Analysis of the Embryological Research Debate in Great Britain and Germany. *Social Politics* 10 (1), 86–128.

Riedel, Ulrike (2004). Das geplante Gentestgesetz. Darstellung der Rechtsprobleme und des grundlegenden rechtlichen Regelungsbedarfs und der Struktur eines Gesetzes zur Regelung von Gentests. Paper presented at the *Ringvorlesung, Bio-GUM,* January 27, in Hamburg, Germany.

Rochefort, David A. and Roger W. Cobb (1994). Problem Definition: An Emerging Perspective. In David A. Rochefort and Roger W. Cobb (eds.) (1994). *The Politics of Problem Definition. Shaping the Political Agenda.* Lawrence, Ks.: University Press of Kansas, 1–31.

Roslin Institute and PPL Therapeutics (1997). Memorandum submitted by the Roslin Institute and PPL Therapeutics (5 March 1997) (CLE 1). In STC (ed.) (1997b). *Fifth Report of Sessions 1996–1997—The Cloning of Animals from Adult*

Cells—Vol. II Minutes of Evidence and Appendices. London: The Stationary Office, 11–12.

Royal Society (1985). *Public Understanding of Science.* London: Royal Society.

— (2004). *Science in Society Report.* London: Royal Society.

Sahm, Stephan (2001). Der doppelte Markl. Ein Fall von Anpassungsfähigkeit. In Christian Geyer (ed.) (2001). *Biopolitik. Die Positionen.* Frankfurt a.M.: Edition Suhrkamp, 200–205.

Salter, Brian and Mavis Jones (2003). The Governance of Human Genetics: Policy Discourse and Construction of Public Trust. *New Genetics and Society* 22 (1), 21–41.

Saretzki, Thomas (2003). Aufklärung, Beteiligung und Kritik: Die 'argumentative Wende' in der Policy-Analyse. In Nils C. Bandelow and Klaus Schubert (eds.) (2003). *Lehrbuch der Politikfeldanalyse.* Munich: Oldenbourg, 391–417.

Schirrmacher, Frank (2001). Bürger Markl. In Christian Geyer (ed.) (2001). *Biopolitik. Die Positionen.* Frankfurt a.M.: Suhrkamp Verlag, 194–199.

Schröder, Gerhard (2001). "Die Notwendigkeit der Abwägung stellt sich immer wieder neu." Gespräch mit Gerhard Schröder. In Christian Geyer (ed.) (2001). *Biopolitik. Die Positionen.* Frankfurt a.M.: Edition Suhrkamp, 88–101.

Sexton, Sarah (1999). If Cloning is the Answer, What was the Question? Power and Decision-Making in the Geneticisation of Health. *The CornerHouse Briefing* No. 16/1999. May 20, 2002, http://www.thecornerhouse.org.uk/item.shtml?x=51972.

— (2000). *How to Talk About Cloning Without Talking About Cloning. Public Discourse in the UK.* Oct 29, 2007, http://www.thecornerhouse.org.uk/item.shtml?x=52016.

— (2001). What's in a Name? The Language and Discourse of Human Embryo Cloning. Paper presented at the Conference *Techniques of Reproduction: Media, Life, Discourse*, December 1–2, at the University of Potsdam, Germany.

Singer, Peter (1982). *Practical Ethics.* Cambridge: Cambridge University Press.

Spaemann, Robert (2001). Gezeugt, nicht gemacht. Die verbrauchende Embryonenforschung ist ein Anschlag auf die Menschenwürde. In Christian Geyer (ed.) (2001). *Biopolitik. Die Positionen.* Frankfurt a.M.: Edition Suhrkamp, 41–50.

Spallone, Pat (1989). *Beyond Conception: The New Politics of Reproduction.* Basingstoke: Macmillan Education.

STC (1997a). House of Common's Science and Technology Committee. *Fifth Report of Sessions 1996–1997—The Cloning of Animals from Adult Cells—Vol. I, Report.* London: The Stationary Office.

— (1997b). House of Common's Science and Technology Committee. *Fifth Report of Sessions 1996–1997—The Cloning of Animals from Adult Cells—Vol. II Minutes of Evidence and Appendices.* London: The Stationary Office.

Thomson, J.A. et al. (1998). Embryonic Stem Cell Lines Derived from Human Blastocysts. *Science* 282, 1145–1147.

Tolmein, Oliver (1990a). *Geschätztes Leben: die neue "Euthanasie"-Debatte.* Hamburg: Konkret Literatur Verlag.

— (1990b). Tödlicher "Zeit"-Geist. *konkret* 06 (1990), 20–25.

— (1993). Liberale Reflexe. *konkret* 07 (1993), 48–52.

— (1998). Ein Recht auf fehlerfreie Babys? *Gen-ethischer Informationsdienst GID* 125/126, 57–58.

Torgerson, Douglas (1985). Contextual Orientation in Policy Analysis: The Contribution of Harold D. Lasswell. *Policy Sciences* 18 (1985), 241–261.

— (1986). Between Knowledge and Politics: The Three Faces of Policy Analysis. *Policy Sciences* 19 (1986), 33–59.

Unkelbach, Alexandra (2001). *Vorbereitung und Übernahme staatlicher Entscheidungen durch plural zusammengesetzte Gremien—Empirische und rechtliche Eckdaten des deutschen Gremienwesens auf Bundesebene.* Speyer: Forschungsinstitut für öffentliche Verwaltung.

Warnock, Mary (1984). *Report of the Committee of Inquiry into Human Fertilization and Embryology.* London: HMSO.

Wilkie, Tom and Elizabeth Graham (2001). Power Without Responsibility. Media Portrayals of British Science. In Arlene J. Klotzko (ed.) (2001). *The Cloning Sourcebook.* Oxford: University Press, 135–150.

Williams, Clare, Jenny Kitzinger, and Lesley Henderson (2003). Envisaging the Embryo in Stem Cell Research: Rhetorical Strategies, Discursive Strategies and Media Reporting of the Ethical Debates. *Sociology of Health and Illness* 25 (7), 793–815.

Winnacker, Ernst-Ludwig (2001). Forschung soll die Welt verändern. In Sigrid Graumann(ed.) (2001). *Die Genkontroverse. Grundpositionen.* Freiburg/Basle/Vienna: Herder Spektrum, 115–122.

Wunder, Michael and Therese Neuer-Miebach (1998a). *Bio-Ethik und die Zukunft der Medizin.* Bonn: Psychiatrie-Verlag.

Wunder, Michael and Therese Neuer-Miebach (1998b). Bio-Ethik und die Zukunft der Medizin. Vorwort. In Michael Wunder and Therese Neuer-Miebach (eds.) (1998a). *Bio-Ethik und die Zukunft der Medizin.* Bonn: Psychiatrie-Verlag, 8–15.

Wynne, Brian (1992). Misunderstood Misunderstanding: Social Identities and the Public Uptake of Science. *Public Understanding of Science*, 1 (3), 281–304.

— (1995). Public Understanding of Science. In Sheila Jasanoff, Gerald E. Markle, James C. Petersen, and Trevor Pinch (eds.) (1995). *Handbook of Science and Technology Studies.* London: Sage, 361–388.

— (1996). May the Sheep Safely Graze? A Reflexive View of the Expert-Lay Knowledge Divide. In Scott Lash, Bronislaw Szerszynski, and Brian Wynne

(eds.) (1996). *Risk, Environment and Modernity. Towards a New Ecology*. London: Sage, 44–83.
— (2001). Creating Public Alienation: Expert Cultures of Risk and Ethics on GMOs. *Science as Culture* 10 (4), 445–481.
— (2002). Risk and Environment as Legitimatory Discourses of Technology: Reflexivity Inside Out? *Current Sociology* 59 (3), 459–477.

Other Sources[136]

Arbeitskreis Bioethik Braunschweig, *Open letter to MPs*. May 26, 2005, http://www.fuente.de/bioethics/bs_000405.htm.
AStA Braunschweig University, *Flyer: Denker die die Welt nicht braucht*. May 18, 2007, http://www.forum-bioethik.de/BioethikN1_3_6.html.

Basic Law for the Federal Republic of Germany (GG), Translation Provided by International Nations. April2, 2006, http://www.iuscomp.org/gla/statutes/GG.htm.
BBSRC, *Biotechnology and Biological Sciences Research Council, biotechnology branch of the Institution of Professionals, Manager and Specialists (IPMS), Webpage*. Sep 2, 2003, http://www.bbsrc.ac.uk.
Berliner Morgenpost (May 15, 2001). *Unmut über der Nationalen Ethikrat. Kritik in Enquête-Kommission des Bundestages*. [Resentment Against the National Ethics Council. Critique from the parliamentary Enquête Commission].
Berliner Zeitung (Jan 19, 2001). *Embryonenschutz ist gesichert. Bundesjustizministerin tritt Befürchtungen eines Kurswechsels entgegen*. [Embryo Protection Secured. Federal Minister of Justice Counters Fears of Change in Policy].
Berliner Zeitung (May 27, 2007). *Aus dem Ausland die Forschung – aus Deutschland die Ethik*, by Jutta Kramm. [Research from Abroad – Ethics from Germany].
BioSkop, Zeitschrift zur Beobachtung der Biowissenschaften, *Webpage*. Oct 8, 2007, http://www.bioskop.de.
BMBF (PM 7/2001). *Bundesministerium für Bildung und Forschung, Press release of 24 Jan. 2001*. Jul 25, 2006, http://www.bmbf.de/press/305.php.
BMG (312-4080/17). *Positionspapier der Bundesministerin für Gesundheit zum geplanten Fortpflanzungsmedizingesetz*. Bonn: Bundesministerium für Gesundheit.
BT DR (10/1581). *Beschlußempfehlung und Bericht des Ausschusses für Forschung und Technologie (18. Auschuß) Einsetzung einer Enquête-Kommission 'Gentechnologie' (Drucksache 10/1353), Einsetzung einer Enquête-Kommission 'Gen-Technik' (Drucksache 10/1388), 07 Juni 1984*. Oct 29, 2007, http://dip.bundestag.de/btd/10/015/1001581.pdf.

[136] Includes parliamentary proceedings and motions, interviews, webpages, statutory instruments, laws, statements, newspaper articles, grey literature, unpublished texts.

BT DR (14/3011). *Antrag auf Initiative der Fraktionen der SPD, CDU/CSU, Bündnis90/Die Grünen, FDP: Einsetzung einer Enquête-Kommission "Recht und Ethik der modernen Medizin", 22 März, 2000*, Oct 29, 2007, http://dip.bundestag.de/btd/14/030/1403011.pdf.

BT DR (14/4098). *Antrag der Abgeordneten Detlef Parr u.a. und der Fraktion der F.D.P.: Präimplantationsdiagnostik rechtliche absichern, 15 September 2000.* Oct 29, 2007, http://dip.bundestag.de/btd/14/040/1404098.pdf.

BT DR (14/6551). *Antrag der SPD und BÜNDNIS 90/DIE GRÜNEN: Für eine umfassende Prüfung des Imports und der Forschung mit embryonalen Stammzellen, 4 Juli 2001.* Oct 29, 2007, http://dip.bundestag.de/btd/14/065/1406551.pdf.

BT DR (14/8101). *Antrag der Abgeordneten Dr. Wolfgang Wodarg. Dr. Hermann Kues, Monika Knoche, Jochen Borchert, Wolfgang Thierse, Ilse Falk u.a. Schutz der Menschenwürde angesichts der biomedizinischen Möglichkeiten—Kein Import embryonaler Stammzellen, 29 Januar 2002.* Oct 29, 2007, http://dip.bundestag.de/btd/14/081/1408101.pdf.

BT DR (14/8102). *Antrag der Abgeordneten Dr. Maria Böhmer, Margot v. Renesse, Andrea Fischer (Berlin), Horst Seehofer, Wolf-Michael Catenhusen et. al. Keine verbrauchende Embryonenforschung: Import humaner embryonaler Stammzellen grundsätzlich verbieten und nur unter engen Voraussetzungen zulassen, 29 Januar 2002.* Oct 29, 2007, http://dip.bundestag.de/btd/14/081/1408102.pdf.

BT DR (14/8103). *Antrag der Abgeordneten Ulrike Flach, Katharina Reiche, Peter Hintze, Dr. Wolfgang Gerhardt u.a. Verantwortungsbewusste Forschung an embryonalen Stammzellen für eine ethisch hochwertige Medizin, 29 Januar 2002.* Oct 29, 2007, http://dip.bundestag.de/btd/14/081/1408103.pdf.

Bürger gegen Bioethik, *Webpage, Ursel Fuchs für Bürger gegen Bioethik (Oct 1, 1997): "Bioethik-Konvention des Europarates auf der deutschen 'Zielgeraden'".* Feb 8, 2007, http://www.fuente.de/bioethik.

Catenhusen, Wolf-Michael (June 29, 1999). *Brief an die SPD Fraktion*, not publicly available.

— (2001). *Interview am Morgen. Zum Streit um die neue Gen-Politik.* In *Deutschlandfunk*, Jan 19, 2001, 7.15 a.m. Radio Interview.

Caesar, Peter (1999). *Präimplantationsdiagnostik. Thesen zu den medizinischen, rechtlichen und ethischen Problemstellungen. Bericht der Bioethik-Kommission des Landes Rheinland-Pfalz vom 20. Juni 1999.* Oct 30, 2007, http://cms.justiz.rlp.de/systemstatic/justiz/pdf/Bioethik.pdf.

Chief Medical Officer, *Letter on the Expert Advisory Group on Therapeutic Cloning in Humans.* Feb 22, 2003 http://www.doh.gov.uk/cegc/cmolet.htm

Chief Medical Officer's Expert Advisory Group on Therapeutic Cloning in Humans, *List of Members.* Feb 22, 2003, http://www.doh.gov.uk/cegc/members.htm.

Chief Medical Officer's Expert Advisory Group on Therapeutic Cloning in Humans, *Terms of reference.* Feb 22, 2003 http://www.doh.gov.uk/cegc/press1.htm.

Daily Mail (Feb 24, 97). *Age of Man-Made Monsters?*, commentary by Paul Johnson.

Daily Telegraph (Feb 28, 97). *Santer Raises Concerns over Human Cloning*, by Roger Highfield.

Daily Telegraph (Feb 26, 1997). *Legal 'Loophole' for Human Clone*, by Roger Highfield.

Daily Telegraph (Mar 1, 1997). *Dolly Scientists to lose Funding*, by Roger Highfield.

Daily Telegraph (Mar 4, 2000.) *Human Cloning is Wrong.*

Daily Telegraph (Mar 5, 1997). *Dolly's Already here—Ethics will have to Cope*, comment by Steve Jones.

Daily Telegraph (Mar 6, 1997). *Warning over Ban on Clone Research.*

Deutsche Gesellschaft für Humangenetik e.V., *Webpage.* Jan 1, 2007, http://www.gfhev.de.

Deutscher Ärztinnenbund (2000). *Stellungnahme zur Präimplantationsdiagnostik (PID).* Jan 20, 2007, http://www.aerztinnenbund.de/Praeimplantationsdiagnostik.0.235.1.html.

Deutsches Ärzteblatt (Dec 10, 1999). *Ethische Grenzfragen in der Medizin. Bundesärztekammer strebt öffentlichen Diskurs an*, by Dr Thomas Gerst. [Ethical Borderline Issues in Medicine. German Medical Association Strives for Public Discourse].

DFG (1999). Deutsche Forschungsgemeinschaft. *Stellungnahme der Deutschen Forschungsgemeinschaft zum Problemkreis "Humane embryonale Stammzellen".* Mar 20, 2000, http://www.dfg.de/aktuelles_presse/reden_stellungnahmen/archiv/download/eszell_d_99.pdf.

Enquête-Kommission "Recht und Ethik der modernen Medizin", *Webpage.* Oct 10, 2004, http://www.bundestag.de/parlament/gremien/kommissionen/archiv14/medi/medi_ext.html.

EschG (1990). *Act for the Protection of Embryos (The Embryo Protection Act) of 13th December 1990.* Federal Lay Gazette, Part I, No 69, issued in Bonn, 19th December 1990, page 2746. July 27, 2006, http://www.bmj.bund.de/files/fe5fef9d2041e75ec37796d8e517288e/1147/ESchG%20englisch.pdf.

Ethik-Beirat (2000). Ethik-Beirat beim Bundesministerium der Gesundheit. *Prädiktive Gentests. Eckpunkte für eine ethische und rechtliche Orientierung. November 2000.* Unpublished Statement of the Ethics Board within the Federal Ministry of Health.

European Bioethics-Convention (1997). *Convention for the Protection of Human Rights and Dignity of the Human Being with Regard to the Application of Biology and Medicine: Convention on Human Rights and Biomedicine.* May 20, 2008 http://conventions.coe.int/Treaty/en/Treaties/Html/164.htm.

Evening Standard (Feb 28, 1997). *The Folly of Dolly. Nobel Peace Prize Winner Professor Joseph Rotblat talking to Suzie Mackenzie.*

Financial Times (Mar 6, 1997). *Baaad Baaan.*

Focus (Jan 15, 2001). *Philosoph ohne Scheuklappen. Kulturstaatsminister Julian Nida-Rümelin entfacht mit Thesen über die Menschenwürde von Embryonen einen Streit im Sinne des Kanzlers.* [Philosopher Who doesn't Wear Blinders: Minister of Cultural Affairs Julian Nida-Rümelin Ignites with Theses on the Human Dignity of Embryos an Argument in Line with the Chancellor's Views].

Forum Bioethik, *Webpage.* May 20, 2007, http://www.forum-bioethik.de.

Frankfurter Allgemeine Zeitung (Jan 26, 2001). *Regierung nicht generell gegen PID.* [Government Not Generally Against PGD].

Frankfurter Allgemeine Zeitung (May 3, 2001). *Die Notwendigkeit der Abwägung stellt sich immer wieder neu. Interview des Bundeskanzlers zur Bioethik* [The Need for Balancing must be Continually Reconsidered. Interview with the Chancellor on Bioethics].

Frankfurter Allgemeine Zeitung (May 15, 2001). *Legt offen was euch bindet. Fragen, die sich der der Ethikrat gefallen lassen muss.* [Reveal the Ties that Bind You. Questions the Ethics Council has to Face].

Frankfurter Rundschau (Jan 16, 2001). *Kurswechsel in der Biotechnologie geplant. Neue Ministerin sieht Gentechnik als "Zukunftsthema".* [Change of Biotechnology Policy Intended. New Minister Sees Genetic Technology as an "Issue of the Future"],

Frankfurter Rundschau (Feb 5, 2001). *Regierung legt Gesetze zum Menschen-Klonen auf Eis.* [Government Puts Human Cloning Law on Hold].

Frankfurter Rundschau (Feb 26, 1997). *Hello Dolly, du Schaf aus der schönen neuen Welt. Erstmals klonten Forscher ein ausgewachsenes Säugetier und fachen die Diskussion neu an: Was darf der Mensch?*, by Michel Emmrich. [Hello Dolly, Sheep from the Brave New World. For the First Time, Researchers Clone a Mature Mammal and Re-kindle the Discussion: What is Allowable for Humankind?].

Frankfurter Rundschau (May 30, 2000). *Die Frau als Reagenzglas. Menschenwürde und Menschenproduktion*, by Kathrin Braun. [Woman as a Petri Dish. Human Dignity and the Production of Humans].

Freitag 23 (June 2, 2000). *Embryo in No-Women's Land*, by Ulrike Baureithel.

Gen-ethisches Netzwerk e.V., *Webpage.* Mar 6, 2005, http://www.gen-ethisches-netzwerk.de.

Genetic Interest Group, *Webpage.* Oct 29, 2003, http://www.gig.org.uk/members.htm.

German National Ethics Council, *Draft Decree Submitted to Cabinet on 25 April 2001, Establishment of a National Ethics Council.* Feb 23, 2006, http://www. ethikrat.de/_english/about_us/decree.html.

German National Ethics Council, *List of Members.* June 20, 2003 http://www.ethikrat.org.

German National Ethics Council, *Press Release 01/2001*. May 11, 2002, http://www.ethikrat.org/presse/pm/200101.html.

German National Ethics Council, *Rules of Procedure*. Feb 23, 2006, http://www.ethikrat.org/_english/about_us/rules_of_procedure.html.

German National Ethics Council, *Webpage*. Feb 23, 2006, http://www.ethikrat.org.

The Guardian (Feb 26, 1997). *Scientist 'able to Create Human Clone'*, by John Arlidge.

HC (Nov 17, 2000). *Hansard House of Commons Debates, 17 Nov. 2000*. Aug 10, 2007, http://www.publications.parliament.uk/pa/cm199900/cmhansrd/vo001117/ebindx/01117-x.htm.

HC (Dec 15, 2000). *Hansard House of Commons Debates, 15 Dec 2000*. Aug 10, 2007, http://www.publications.parliament.uk/pa/cm200001/cmhansrd/vo001215/debindx/01215-x.htm.

HC (Dec 19, 2000). *Hansard House of Commons Debates, 19 Dec. 2000*. Aug 10, 2007, http://www.publications.parliament.uk/pa/cm200001/cmhansrd/vo001219/ebindx/01219-x.htm.

HFE Act (1990). *Human Fertilisation and Embryology Act 1990*. Jan 15, 2007, http://www.opsi.gov.uk/acts/acts1990/Ukpga_19900037_en_1.htm.

Human Fertilisation and Embryology Authority, *Webpage: About Us*. Oct 29, 2003, http://www.hfea.gov.uk/AboutHFEA.

Human Fertilisation and Embryology Authority, *Code of Practice I*. July 18, 2008, http://www.hfea.gov.uk/docs/1st_Edition_Code_of_Practice_1991.pdf.

Human Fertilisation and Embryology Authority, *Webpage*. Aug 10, 2007, http://www.hfea.gov.uk.

HFE (Research Purposes) Act (2001). *The Human Fertilisation and Embryology (Research Purposes) Regulations 2001, Statutory Instrument 2001, No. 188*. April15, 2003, http://www.opsi.gov.uk/si/si2001/20010188.htm.

HGAC (P/97/180). *Human Genetics Advisory Commission Press Notice: New Genetics Advisory Commission put Public Confidence on Top of its Agenda*. Oct 20, 2007, http://www.advisorybodies.doh.gov.uk/hgac/index.html.

Human Genetics Advisory Commission, *Webpage*. Oct 20, 2007, http://www.advisorybodies.doh.gov.uk/hgac/index.html.

HL (Jan 22, 2001). *Hansard House of Lords Debates, 22 Jan 2001*. Aug 10, 2007, http://www.publications.parliament.uk/pa/ld200001/ldhansrd/vo010122/text/10122-04.htm#10122-04_head2.

The Independent (Feb 28, 1997). *Cloning Presents an Opportunity, not a Threat*.

Institut Mensch Ethik und Wissenschaften, *Webpage*. Jan 10, 2007, http://www.imew.de.

Institut Mensch Ethik und Wissenschaften, *About Us*. Jan 10, 2007, http://www.imew.de/imew.php/cat/74/title/About_the_Institute.

Institut Mensch Ethik und Wissenschaften, *Decree on the Establishment of the IMEW.* Jan 10, 2007, http://www.imew.de/imew.php/cat/78/title/Gruendungserklaerung.

Interessengemeinschaft Kritische Bioethik Deutschland, *Webpage on the European Bioethics Convention.* April 1, 2005, http://www.bioethik-konvention.de.

Interview D1 (Oct 21, 2005). *Interview with the Chair of the Parliamentary Inquiry Commission 'Chances and Risks of Genetic Technology' (1984-1986)*, held on 21 Oct 2005 in Berlin.

Interview D2 (Oct 27, 2005). *Interview with the Chair of the German National Ethics Council (2001–2005), Member of the National Ethics Board (2001–2007)*, held on 27 Oct 2005 in Berlin.

Interview D4 (Nov 10, 2005). *Interview with Andrea Fischer (Bündnis 90/Die Grünen), Federal Health Minister (1998–2001),* held on 10 Nov 2005 in Berlin.

Interview D5 (Nov 30, 2005). *Interview with a Member of the Parliamentary Inquiry-Commission 'Law and Ethics of Modern Medicine' (2000–2002), Co-initiator of the Parliamentary 'Alliance for Human Dignity',* held on 30 Nov 2005 in Berlin.

Interview D 6 (Dec 1, 2005). *Interview with Ernst Benda, Chair of the Benda Commission (1984–1985), President of the Constitutional Court (retired)*, held on 01 Dec 2005 in Berlin.

Interview D7 (Jan 18, 2006). *Interview with a Member of the Ethics Board in the Health Ministry (1999–2001), Member of the German National Ethics Council (2001–2007),* held on Jan 18, 2006 in Hamburg.

Kölner Stadtanzeiger (May 17, 2001). *Deutlicher Mangel an Aufrichtigkeit.* [A Clear Lack of Sincerity].

McEldoney, John. *Memorandum presented to the House of Lords Select Committee on the British Constitution.* Aug 10, 2007, http://www.parliament.the-stationery-office.co.uk/pa/ld200203/ldselect/ldconst/168/16809.htm.

Netzwerk (2001). Netzwerk gegen Selektion durch Pränataldiagnostik. *Die aktuelle Gentechnik-Debatte, PID, Beratung.* Rundbrief 12.

— (2006). Netzwerk gegen Selektion durch Pränataldiagnostik, *Stellungnahme zu Präimplantationsdiagnostik.* Jan 1, 2007, http//www.bvkm.de/1-10/praenataldiagnostik,netzwerk,stellungnahme.

The Observer (Feb 23, 1997). *Scientists Clone Adult Sheep*, by Robin McKie.

The Observer Review (Mar 10, 1996) *A Clone Again, Naturally*, by Robin McKie.

PLP (14/173). *Plenarprotokoll 14/173, Deutscher Bundestag, Stenographischer Bericht, 173. Sitzung, Berlin, 31 May* 2001. Oct 29, 2007, http://dip.bundestag.de/btp/14/14173.pdf.

PPL Therapeutics Ltd, *Webpage*. Jan 20, 2006, http://www.ppl-therapeutics.co.uk/welcome/welcome.html.

Rheinischer Merkur (Jan 19,2001). *Wir streiten für die Würde*, von Matthias Gierth. [We Fight for Dignity].

Royal Society, *Webpage*. Mar 10, 2007, http://royalsoc.ac.uk.

SchKG (1992). *Gesetz zur Vermeidung und Bewältigung von Schwagerschaftskonflikten*. Dec 13, 2006, http://bundesrecht.juris.de/beratungsg/index.html.

Schmidt, Ulla (Jan 31, 2001). *Gesundheitspolitik des Vertrauens. Erstes Pressegespräch mit der Bundesgesundheitsministerin Ulla Schmidt*. [A Trustworthy Health Policy. First Press Conference with the Federal Minister of Health Ulla Schmidt]. Feb 1, 2005, http://www.bmg.bund.de/deu/gra/aktuelles/reden/bmgs/index_2763.cfm.

Schröder, Gerhardt (April 5, 2001). *Rede bei der Katholischen Akademie, 5 April 2001*. Jan 12, 2002, http://www.bundesregierung.de/Reden-Interviews/Reden-,11636.35770/rede/Rede-von-Bundeskanzler-Gerhard.htm.

— (July 4, 2001). Rede von Bundeskanzler Gerhard Schröder auf der Jahresversammlung der Deutschen Forschungsgemeinschaft (DFG) am 4 Juli 2001 in Berlin, *Bulletin*, 46-1 v. 5 Juli 2001/BK—Annual Meeting of the German Research Foundation.

Der Spiegel (May 21, 2001). *Debatte nicht verkürzen. Bundeskanzler Gerhard Schröder, 57, zur Berliner Rede von Bundspräsident Johannes Rau und über moralische Argumente für Gentechnik und Biomedizin*, [Don't Curtail the Debate. Chancellor Gerhard Schröder, 57, on the Berlin Speech held by President Johannes Rau and on Moral Arguments for Genetic Technology and Biomedicine].

StGB (1998). *Strafgesetzbuch*. Dec 13, 2006, http://bundesrecht.juris.de/stgb/index.html.

StZG (2002). *Act Ensuring Protection of Embryos in Connection with the Importation and Utilization of Embryonic Stem Cells (The Stem Cell Act) of 28 June 2002 (unofficial translation)*. Feb 8, 2007, http://www.bundestag.de/parlament/gremien/kommissionen/archiv15/ethik_med/archiv/stammzellgesetz_engl.pdf.

Süddeutsche Zeitung (Mar 25/26, 2000). *Die SPD-Abgeordnete war ursprünglich eine Gegnerin jener Kommission, für deren Vorsitz sie nun vorgeschlagen wird*. [SPD MP was Originally Against the Commission of which she has now been Appointed Chair].

Südwest Presse Ulm (Feb 3, 2000). *Menschen dürfen nicht zu Objekten werden*. [Humans Must Not Become Objects], Interview with Heinz Krus, spokesman of the citizen initiative against the Bioethics Convention. Dec 11, 2006, http://www.fuente.de/bioethik/swp_000203.htm.

Survey on the National Regulations in the European Union Regarding Research on Human Embryos. April 4, 2002, http://europa.eu.int/comm/european_group_ethics/docs/nat_reg.pdf.

Der Tagesspiegel (Jan 3, 2001). *Wo die Menschenwürde beginnt. Das Klonen von Embryonen: ein Heilsweg oder der Anfang eines gespenstischen Menschenbilds? Auch Deutschland kann einer neuen bio-ethischen Debatte nicht mehr ausweichen*, by Julian Nida-Rümelin. [Where Human Dignity Begins. The Cloning of Embryos: A Way to Healing or the Advent of an Eldritch Image of Human Beings? Germany Cannot Avoid a New Bioethics Debate].

Der Tagesspiegel (Jan 25, 2001). *Rot-Grün will Einsatz von Gentechnik nun doch erleichtern.* [Red-Green Coalition Wants to Alleviate Genetic Technology Application].

Der Tagesspiegel (May 15, 2001). *Außer Konkurrenz. Enquetekommission pocht bei Gentechnik auf ihre Zuständigkeit.* [Non-Competitive. The Enquête Commission Insists on its Authority in Genetic Issues.]

die tageszeitung (Feb 13, 2001). *Gen tut alles. Forschungsministerin Bulmahn steckt 870 Millionen Mark in die Erforschung des menschlichen Erbguts. Schröder will zur Sicherheit einen Nationalen Ethikrat einberufen.* [The Gene can accomplish all. Research Minister Bulmahn Invests 870 Million Deutsch Marks in Research on Human Genetic Information. To Guarantee Against Abuse Schröder Intends to Set up a National Ethics Council].

die tageszeitung (April 12, 2000). *Biomedizin statt sozialer Politik. Reagenzglasbefruchtung, vorgeburtliche Gendiagnostik und Präimplantationsdiagnostik—Sind das die neuen Instrumente zur Befreiung der Frauen?*, by Sigrid Graumann. [Biomedicine Instead of Social Politics. Test Tube Fertilisation, Prenatal Diagnosis, and Pre-implantation Genetic Diagnosis—Are These the New Instruments for Liberating Women?].

The Times (Feb 27, 1997). *Scientists "Close to Recreating the Dead",* by Simon de Bruxelles.

The Times (Mar 6, 1997). *Embryo Watchdog Backs Trials in Human Cloning*, by Jeremy Laurence

The Times (Mar 8, 1996a). *Cloning Breakthrough Sounds Ethical Alarm. Experts Fear Experiment Brings Fiction of the Mad Scientist One Step Closer to Reality*, by Nigel Hawks.

The Times (Mar 8, 1996b). *We are Clones of Dr Faust. Roger Scruton on the Descent from Genesis to Genetics.*

Tübinger Initiative gegen die geplante Bioethik-Konvention, *Webpage.* Feb 7, 2007, http://www.fuente.de/bioethik/.

VDW (1996). Ad-hoc Gruppe der Vereinigung Deutscher Wissenschaftler (VDW). *Forschungsfreiheit öffentlich verantworten. Stellungnahme zur DFG-Denkschrift: "Forschungsfreiheit. Ein Plädoyer für bessere Rahmenbedingungen der Forschung in Deutschland."* May 30, 2007, http://www.vdw-ev.de/archiv/stellung-Forschungsfreiheit.pdf.

Vereinigung Deutscher Wissenschaftler, *Webpage*. May 30, 2007, http://www.vdw-ev.de.

Die Welt (Mar 9, 2001). *Severino Antinori will der Schöpfer des ersten menschlichen Klons sein.* [Severino Antinori Wants to be the Creator of the First Human Clone].

Die Welt (May 18, 2001). *Ein Kanzleramts-Ethiker mit Bedenken.* [A Concerned Chancellery's Ethicist].

Die Woche (Dec 20, 2000). *Der neue Mensch. Beitrag zur Gentechnik von Bundeskanzler Gerhard Schröder für die Wochenzeitung Die Woche.* [The New Human Being. An Article on Gene Technology by Chancellor Gerhard Schröder for the Weekly Newspaper Die Woche].

Die Zeit (Sep 16, 1999). *Regeln für den Menschenpark. Ein Antwortschreiben zum Brief über den Humanismus—die Elmauer Rede*, by Peter Sloterdijk. [Rules for the Human Zoo. A Reply to the Letter on Humanism—the Elmau Lecture].

Die Zeit (Dec 28, 2000). *Das Jahr der Biopolitik*, by Gero v. Randow. [The Year of Biopolitics].

Die Zeit (No 24, 2001). *Heizer in der Bimmelbahn. Wie Wolfgang Clement den Bioethikern Dampf macht,* by Hans Schuh. [Stoker in the Slow Train. How Wolfgang Clement Puts Pressure on Bioethicists].

Die Zeit (June 13, 2001). *Die Abtreibungsfalle*, by Reinhard Merkel. [The Abortion Trap].

Die Zeit, *Webpage*. Dec 30, 2005, http://www.zeit.de.

Zentrale Ethik Kommission zur Stammzellforschung, *Statutory Instrument*. Oct 19, 2006, http://www.bgblportal.de/BGBL/bgbl1f/bgbl102s2663.pdf.

Zentrale Ethik Kommission zur Stammzellforschung, *Webpage*. Oct 19, 2006, http://www.rki.de/cln_048/nn_207098/DE/Content/Gesund/Stammzellen/ZES/zes__node.html?__nnn=true.